About the Author

JOSEPH ROMM is the founder and executive director of the Center for Energy and Climate Solutions and a fellow at the Center for American Progress. Under President Clinton he was acting assistant secretary at the Department of Energy, heading the Office of Energy Efficiency and Renewable Energy. Author of the award-winning *The Hype About Hydrogen*, he holds a Ph.D. in physics from MIT and lives in Washington, D.C.

HELL AND HIGH WATER

THE GLOBAL WARMING SOLUTION

Joseph Romm

HARPER PERENNIAL

NEW YORK • LONDON • TORONTO • SYDNEY

HARPER ● PERENNIAL

FIRST HARPER PERENNIAL EDITION PUBLISHED 2008.

Designed by Daniel Lagin

Library of Congress Cataloging-in-Publication Data has been applied for.

ISBN: 978-0-06-117213-7 (pbk.)

08 09 10 11 12 DIX/RRD 10 9 8 7 6 5 4 3 2 1

To Patricia and the hope she brings

The rarest attribute in any society and culture when things are going generally well . . . is to notice certain cracks in the edifice, some defects and problems, which if not attended to could in time undermine the happy ambience and bring on distress and terror.

—Norman F. Cantor, *In the Wake of the Plague*, 2001

History does not forgive us our national mistakes because they are explicable in terms of our domestic politics.

—George Kennan, 1950

CONTENTS

HELL AND HIGH WATER

INTRODUCTION

We are on the precipice of climate system tipping points beyond which there is no redemption.
   —James Hansen, director, Goddard Institute for
   Space Studies (NASA), December 2005

The ice sheets seem to be shrinking 100 years ahead of schedule.
   —Richard Alley, Penn State climate scientist,
   May 2006

Imagine if inland United States were 10°F hotter, with many states ravaged by mega-droughts and the widespread wildfires that result. At the same time, our coasts were drowning from a 5- to 10-foot increase in sea levels, which were relentlessly climbing 5 to 10 inches a decade or more toward an ultimate sea-level rise of *80 feet.*

   This "Hell and High Water" scenario is not our certain future, but it is as likely as the bird flu pandemic we are feverishly fighting to fend off. And it could come as soon as the second half of this century, given the many early warning signs of accelerated climate change that scientists have spotted.

Long before then, the temperature of the inland United States will be rising nearly 1°F *per decade,* enough to cause continual heat waves and searing droughts. At the same time, sea levels will be rising a few inches every decade, with much of our Atlantic and Gulf coasts battered year after year after year by super-hurricanes with savage storm surges.

Let's call this phase Planetary Purgatory, when the world comes to know that 20-foot sea-level rise is all but inevitable, and we must endure a desperate multidecade ordeal to correct the mistakes of the past, to keep sea-level rise as low and slow as possible—to avoid the full fury of Hell and High Water. If the politics of inaction and delay that have triumphed in this country continues for another decade, then Planetary Purgatory is *the likely future facing our country before midcentury—probably in your own lifetime.*

According to a March 2006 Gallup Poll, only about a third of Americans understand that global warming will "pose a serious threat to you or your way of life in your lifetime." And if you think that global warming will mainly affect other, poorer countries, or that we can delay acting until we have new technologies, you come by your opinions honestly. Many of the most sophisticated policy makers and journalists also just don't get it—they don't understand how global warming will ruin America for the next fifty generations if we don't act quickly.

The widespread confusion about our climate crisis is no accident. For more than a decade, those who deny that climate change is an urgent problem have sought to delay action on global warming by running a brilliant rhetorical campaign and spreading multiple myths that misinform debate. As a result, many people still believe global warming is nothing more than a natural climate cycle that humans cannot influence, or that it might even have positive benefits for this nation. Neither is true. The science is crystal clear: We humans are the primary cause of global warming, and we face a bleak future if we fail to act quickly.

We must reverse the growth in U.S. greenhouse gas emissions and assert leadership to bring every country, especially China, along with us. We created this problem, and we can solve it.

I have spent nearly two decades working to achieve this clean-energy future. The cost is far lower and the benefits far higher than the opponents of action argue, yet the winning solution is not well understood by policy makers, the media, or the public. As I will spell out, a very few states, most notably California, have embraced critical clean-energy strategies; we need to adopt these nationwide. These strategies would also deliver vast benefits—a sharp drop in foreign-oil imports and in our massive trade deficit, a large gain in air quality and health, and a big boost in high-wage, high-tech jobs. This win-win-win future, however, requires a rapid change in both domestic and foreign policy. Congress and the president would have to join together to embrace the aggressive government-led regulatory and technological strategy that they have repeatedly failed to adopt.

Time is short. *We have at most a decade to sharply reverse course.*

If we fail to act in time, global warming will profoundly and irreversibly remake every aspect of American life—where we live, how we live, how we grow food and how much we grow, what and how we drive, how we relate to other countries, and so on.

As catastrophic sea-level rise becomes inevitable, we will become consumed by urban triage—how to decide which major seaside cities can be saved and which cannot. Every seaside city will be threatened: Seattle, San Francisco, Los Angeles, San Diego, Houston, New Orleans, Mobile, St. Petersburg, Miami, Jacksonville, Savannah, Norfolk, Baltimore, Jersey City, New York City, New Haven, Providence, Boston, and Portland, Maine—along with hundreds of smaller cities. No other nation has as much wealth along its shores.

The most devastating flooding probably won't occur until after 2100, but long before then, painful choices will be forced upon the nation over and over again by record-breaking hurricane seasons with ever more devastating storm surges. *Protecting* dozens of major coastal cities from flooding will be challenging enough; *rebuilding* major coastal cities destroyed by super-hurricanes will be an overwhelming task, especially if we allow rapid sea-level rise to become unstoppable in the second half of the century. We do not appear to be willing to spend the money now to protect New Orleans from another hurricane like Katrina—let alone the combination of such a hurricane plus the coming sea-level rise—and that suggests that the city will not survive the next super-hurricane, which is likely to come within the next few decades.

In the political realm, global warming is poised to become the Achilles' heel of the American conservative movement. Its fatal mistake: turning global warming into a partisan ideological issue. Conservative politicians, pundits, and think tanks are staking their movement's future on denying the science and delaying the solution. But while they can stop the nation from acting to prevent the worst of global warming, they cannot stop the searing reality of their perverse blunders from becoming painfully clear to all.

The former chair of the Senate Environment and Public Works Committee, Senator James Inhofe, calls global warming "the greatest hoax ever perpetrated on the American people" and holds hearings where he and witnesses such as novelist Michael Crichton belittle the work of the entire scientific community. President George W. Bush has blocked all national efforts to limit greenhouse gas emissions and has thwarted international efforts to develop stronger emissions controls. If they continue on this course, Bush and Inhofe will go down in history with other leaders such as Herbert Hoover and Neville Chamberlain who were blind to their nation's gravest threats.

Imagine the impact catastrophic climate change will have internationally. For decades, the United States has been the moral, economic, and military leader of the free world. What will happen when we end up in Planetary Purgatory, facing 20 or more feet of sea-level rise, and the rest of the world blames our inaction and obstructionism, blames the wealthiest nation on earth for refusing to embrace even cost-effective solutions that could spare the planet from millennia of misery? The indispensable nation will become a global pariah.

Predicting the unpredictable and imagining the unimaginable consequences of this climate crisis are among my major goals here. Anyone who wants to understand the disastrous but largely avoidable fate to which we are committing America and the rest of the world—as well as the only sensible way to avoid catastrophe—must understand the three driving forces: climate science, energy trends and technology, and global-warming politics. This book is a primer on all three.

I first became interested in global warming in the mid-1980s, studying for my physics Ph.D. at the Massachusetts Institute of Technology and researching my thesis on oceanography at the Scripps Institution of Oceanography in California. I was privileged to work with Walter Munk, one of the world's top ocean scientists, on advanced acoustic techniques for monitoring temperature changes in the Greenland Sea.

A few years later, as special assistant for international security to Peter Goldmark, president of the Rockefeller Foundation, I found myself listening to some of the nation's top experts on these issues. Even a generation ago, they knew the gravest threats that would face us today. They convinced me that global warming was the most serious long-term, *preventable* threat to the health and well-being of this nation and the world. In the mid-1990s I served for five years in the U.S. Department of Energy. As an acting assistant secretary, I helped develop a climate-technology strategy for the nation. More recently, I have worked with some of the nation's leading corporations, helping them to make greenhouse gas reductions and commitment plans that also handsomely boost their profits.

But the awesome nature of the tragedy we face did not hit home for me until Hurricane Katrina struck my brother and his family. A 30-foot wall of water with waves up to 55 feet high crashed into Pass Christian, Mississippi, where my brother lived with his wife and son. The ferocious storm surge destroyed their house, one mile inland, while they stayed in a Biloxi shelter. This book began as a research effort I started so I could advise my brother on the tough question of whether or not he should rebuild his home.

What I learned is that global warming has already begun making Atlantic hurricanes far more destructive. Energy and moisture picked up from warmer Gulf waters produce more intense winds and rain. And in the case of Katrina, that extra punch may be what destroyed the levees protecting New Orleans—the "straw that breaks the camel's back," in the words of Dr. Kevin Trenberth, head of Climate Analysis at the National Center for Atmospheric Research.

Katrina reveals what is to come for this country. On our current path, all our great Gulf and Atlantic coast cities are at risk of meeting the same fate as New Orleans.

If the situation is so dire, why aren't more people running around with their "hair on fire," as CIA director George Tenet was in the summer of 2001, trying to get someone, anyone, to hear his warnings about an impending terrorist attack? In fact, much of the scientific community has been astonished that their increasingly strong and detailed warnings have been either ignored or attacked. I was astounded to learn the full extent of the Bush administration's methods for muzzling government climate scientists and censoring their work, which has prevented their urgent message from reaching the American public. The highest ranks of the National

Hurricane Center and the National Oceanic and Atmospheric Administration have misinformed the public about the likely danger of—and increased number of—future super-hurricanes.

One reason I wrote this book is to give voice to those scientists whose warnings have gone unheard or unheeded.

Three full decades have passed since the National Academy of Sciences, the nation's most prestigious scientific body, first warned that uncontrolled greenhouse gas emissions might raise global temperatures a staggering 10°F and raise sea levels 20 feet—and yet the nation has still not taken any serious action. In stunning contrast, less than five years after climate scientists warned us in 1974 that chlorofluorocarbons were destroying the earth's ozone layer, America voluntarily banned their use in spray cans, and a decade later President Reagan and Vice President Bush led the way to creating an international treaty banning them.

One key goal of this book is to provide a fuller answer to the puzzle of why this country has failed to act on global warming. As we will see, the failure stems from weaknesses inherent in the scientific community, strategic and messaging mistakes made by environmentalists and progressive politicians, flaws in the media's coverage of science, and an insidious effort to exploit those weaknesses, mistakes, and flaws by conservative political leaders such as President Bush as well as a small group of scientists and conservative think tanks with funding from fossil fuel companies.

Global warming has also proved intractable because this country has refused to adopt a sensible energy strategy. Our political leaders won't even require Detroit to build fuel-efficient vehicles—which would not only reduce greenhouse gas emissions but also save consumers money and cut oil imports during a time of war in the Persian Gulf and record-high gasoline prices at home. Instead, we have squandered many years and hundreds of millions of dollars on a misguided—and, as we will see, largely cynical—technology strategy focused on hydrogen-fuel-cell cars that offers no hope of cutting overall greenhouse gas emissions (or oil imports) until mid-century, if ever.

Energy is a subject with as many myths as climate science. The most destructive one is that we cannot tackle global warming until we develop new breakthrough technologies. In fact, the reverse is true. We have cost-effective technologies today that can sharply reduce global-warming pollution. If we don't start reducing greenhouse gas emissions very soon with the technology we have, it will be too late for something new to do us any good. Interestingly, while the climate scientists I talk to invariably warn about crossing thresholds of greenhouse gas pollution that could bring catastrophe, few know enough about energy issues to fully understand just how little time we have to act. This book attempts to bridge the gap between climate science and energy policy.

The first half focuses on our country's future if we don't reverse course immediately. Front and center are the climate system's deadly feedback loops—the vicious cycles whereby an initial warming causes changes that lead to more and more warming—all of which reduce the time we have available to act.

The second half of this book focuses on the politics and the solution. I examine the brilliant disinformation campaign created to sow doubt about climate science and the equally clever campaign to create confusion about the crucial climate solutions. Then I lay out how we can achieve deep reductions in greenhouse gas emissions in the electricity and transportation sectors without raising the nation's overall energy bill. We'll see what the car and the fuel of the future will be, since it isn't going to be fuel-cell vehicles running on hydrogen. Finally, I explore the role of China and the role of the media.

The main goal of this book is to lay out the climate-change warning clearly and persuasively. My hair *is* on fire. And yours should be, too.

PART I

THE SCIENCE AND THE FUTURE

CHAPTER ONE

THE CLIMATE BEAST

The paleoclimate record shouts out to us that, far from being self-stabilizing, the Earth's climate system is an ornery beast which overreacts even to small nudges.

　　　　　—Wallace Broecker, climate scientist, 1995

The ongoing Arctic warming corresponds to the predictions of the more pessimistic climate models. By extension, the pessimistic scenarios of climate change can be expected to unfold in the rest of the Northern Hemisphere.

　　　　　—Louis Fortier, climate scientist, June 2006

We are on the brink of taking the biggest gamble in human history, one that, if we lose, will transform the lives of the next fifty generations. I will not focus here on the history of how we came to our current understanding of global warming or on the thousands of brilliant scientists whose work brings us this knowledge. That story has been well told already, particularly by Spencer Weart, a physicist and historian, who has put on the web his extensive "hypertext history of how scientists came to (partly) understand what people are doing to cause climate change."

Similarly, I will not lay out more than briefly the scientific underpinnings for our understanding of global warming or of the extensive and conclusive evidence that climate change is occurring. The case has been made again and again by hundreds of top scientists who have done research and analysis for prestigious bodies such as the U.N.'s Intergovernmental Panel on Climate Change (IPCC), the National Academy of Sciences, and the Arctic Council, the nations that border the Arctic Circle, including ours, in its December 2004 *Arctic Climate Impact Assessment.*

How strong is the scientific consensus? Back in 2001, President George W. Bush asked the National Academy of Sciences for a report on climate change and on the conclusions of the IPCC assessments on climate change. The eleven-member blue-ribbon panel, which included experts previously skeptical about global warming, concluded: Temperatures are rising because of human activities; the scientific community agrees that most of the rise in the last half-century is likely due to increased greenhouse gas concentrations in the atmosphere; and "the stated degree of confidence in the IPCC assessment is higher today than it was 10, or even 5 years ago."

Back in 2001, Donald Kennedy, *Science* editor in chief and president emeritus of Stanford University, commented on the steady stream of peer-reviewed reports and articles documenting global climate change appearing in his and other journals: *"Consensus as strong as the one that has developed around this topic is rare in science."* And in December 2004, *Science* published the results of an analysis of nearly a thousand scientific studies appearing in refereed scientific journals between 1993 and 2003. The conclusion:

> This analysis shows that scientists publishing in the peer-reviewed literature agree with IPCC, the National Academy of Sciences, and the public statements of other professional soci-

eties. Politicians, economists, journalists, and others may have the impression of confusion, disagreement, or discord among climate scientists, but that impression is incorrect.

The strong consensus has grown even stronger because the case has grown even stronger. "Evidence of global warming became so overwhelming in 2004 that now the question is: What can we do about it?" That was *Discover* magazine in its January 2005 issue, which called the ever-strengthening case for climate change the top science story of the year.

"There can no longer be genuine doubt that human-made gases are the dominant cause of observed warming," explained James Hansen, director of NASA's Goddard Institute for Space Studies, in April 2005. Hansen led a team of scientists that made "precise measurements of increasing ocean heat content over the past 10 years," which revealed that the earth is absorbing far more heat than it is emitting into space, confirming what earlier computer models had shown about warming. Hansen called this energy imbalance the "smoking gun" of climate change.

In June 2005 the national science academies of the United States, Brazil, Canada, China, France, Germany, India, Italy, Japan, Russia, and the United Kingdom issued a joint statement on climate change urging the nations of the world to take prompt action to reduce greenhouse gas emissions. So far, the world has not listened. Worse, in December 2005, the U.S. government shamelessly blocked the world from acting at an international conference in Montreal that was aimed at developing the next steps for action on climate change.

If you are interested in understanding the detailed evidence for global warming and climate science, if you want to know the answer to key questions such as "How do we know that recent carbon dioxide increases are due to human activities?" or "How do we know that an increase in solar activity is *not* the cause of recent planetary warming?" bookmark the website www.realclimate.org. This site, run by climate experts, answers these and other questions and discusses the latest findings.

My focus instead is the question of the century: *Do we humans have the political will to stop the great ice sheets of Greenland and West Antarctica from melting . . . to stop Hell and High Water?*

PUNCHING THE CLIMATE BEAST

Whether human activity will trigger catastrophic climate change depends on two factors: how much heat-trapping, climate-altering greenhouse gases we pour into the atmosphere, and how the climate system responds to those gases. Recent evidence indicates the climate is more sensitive than had been widely thought. Louis Fortier, Canada Research chair on the Response of Arctic Marine Ecosystems to Climate Change at Université Laval, echoed the thinking of many climate scientists when he said at a June 15, 2006, transatlantic conference that we should now expect the more "pessimistic scenarios" of climate change. Let's try to understand why.

The greenhouse effect has made the life we know possible. The basic physics is straightforward. Our sun pours out intense amounts of visible light, along with radiation, across the electromagnetic spectrum, including ultraviolet and infrared. The sun's peak intensity is in visible light. Of the solar energy hitting the top of the atmosphere, about 30 percent is reflected back into space—by the atmosphere itself (including clouds) and by the earth's surface (land, ocean, and ice). The rest is absorbed, mostly into the earth but some by the atmosphere. This process heats up the planet. The earth reradiates the energy it has absorbed mostly as heat, infrared radiation.

Some naturally occurring atmospheric gases let visible light escape through into space while trapping certain types of infrared radiation. Because these greenhouse gases, including water and carbon dioxide (CO_2), trap some of the reradiated heat, they act as a partial blanket that helps keep the planet about 60°F warmer than it otherwise would be, and that is ideal for us humans.

Since the dawn of the industrial revolution 250 years ago, humankind has been spewing vast quantities of extra greenhouse gases into the atmosphere, causing more and more heat to be trapped. Carbon dioxide released from burning fossil fuels—coal, oil, and natural gas—makes up 85 percent of U.S. greenhouse gas emissions. For most of the past two centuries, few worried about the consequences. But thanks to the work of thousands of scientists, the risks are now clear: We are engaging in a dangerous, planetwide, uncontrolled experiment as these emissions push our climate system into a different state, a far less hospitable state, than human civilization has ever known.

The first remarkable, and ominous, fact about our climate system is that it is not steady, not self-stabilizing. It is an "ornery beast," as climatologist Wallace Broecker calls it. That beast is shaped like a spiked monster. Consider figure 1, the temperature record for the past 400,000 years, derived primarily from Antarctic ice-core data. The mile-long ice cores, drilled by hardy scientists in the harshest of climates, are a record of annual snowfall. The trace gases trapped in the ice layers reveal the temperature and atmospheric composition year by year.

The sawtooth temperature pattern reveals that long ice ages (the valleys in the figure) have been followed by relatively brief, warm interglacial periods (the peaks), such as the one we're in now that began with the end of the last ice age some 10,000 years ago. These ten millennia of mild weather have made possible human civilization as we know it today. And yet, as the figure shows, the interglacial period we now live in is only a very few degrees centigrade warmer than the average temperature of the last ice age, which lasted about 100,000 years.

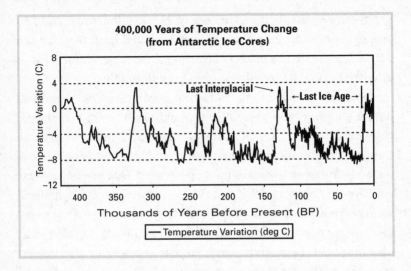

Figure 1. The temperature record for the past 400,000 years from Antarctic ice cores. The "0" for temperature on the *y* axis is the average temperature from 1880 to 1899. The last interglacial (warm period) began 131,000 years ago and lasted 15,000 years. The most recent ice age started about 110,000 years ago and ended about 10,000 years ago. We have warmed about 0.8°C since the industrial revolution (not shown on this chart).

The second ominous fact: Warming can happen fast. As a 2002 study by the National Academy of Sciences explained:

Recent scientific evidence shows that major and widespread climate changes have occurred with startling speed. For example, *roughly half the north Atlantic warming since the last ice age was achieved in only a decade,* and it was accompanied by significant climatic changes across most of the globe. Similar events, including local warming as large as 16°C, occurred repeatedly during the slide into and climb out of the last ice age. [Emphasis added.]

Take a look at figure 1 again. Notice that the warming line at the start of every brief mild interglacial age typically rises very sharply.

You might expect that the temperature would go up gradually. But instead, it all happens very quickly.

Before the 1990s, most scientists saw climate change as a slow, gradual process, linked to variations in Earth's orbit that changed on the timescale of tens of thousands of years, and to changes driven by continental drift over the course of tens of millions of years. But scientific advances, such as those that have allowed us to unlock the evidence found in ancient ice, reveal that huge temperature swings and a doubling of precipitation have occurred "in periods as short as decades to years."

The final ominous fact: The climate has changed most quickly when it has been "forced" to change, such as by increased heating from the sun or from greenhouse gases. We are now forcing the climate to change much, much faster than nature has in the past. The NAS study noted, "Abrupt climate changes were especially common when the climate system was being forced to change most rapidly." The risk, then, is that the rapid greenhouse warming we ourselves are causing today increases the chances for "large, abrupt, and unwelcome regional or global climatic events."

THE FAST FATAL FEEDBACKS

The climate system's ability to warm so rapidly suggests it has strong feedbacks or vicious cycles whereby a small initial warming leads to a disproportionately huge heating. It works like this: Something triggers an initial warming, a forcing event, such as a change in the path Earth takes to orbit the sun, and that brings more intense sunshine (solar insolation) to the planet. Then feedbacks reinforce warming and our planet heats up faster. What kind of vicious cycles? Three in particular are well known.

First, warming causes sea ice to melt and glaciers to retreat. Highly reflective white ice is replaced by the blue sea or dark land, both of which absorb far more solar energy. So the blue oceans and the dark earth heat up more, causing even more ice melting, which results in a larger decrease in Earth's reflectivity (albedo), and that leads to more heating, and so up and up the temperature spiral. We can witness this classic feedback today at the North Pole, where the white summer ice cap has shrunk more than 20 percent from 1978 to 2005, a loss of 200,000 square miles of ice, an area twice the size of Texas, in a single generation.

Second, warming increases evaporation and the amount of water vapor in the air. Water vapor is a greenhouse gas. More water vapor means more warming, which means more water vapor, and so on and on. A 2005 study found that upper-atmospheric moistening from 1982 to 2004 was being accurately modeled by climate scientists. As we will see, more water vapor also leads to more intense hurricanes and rainstorms.

Third, warming can cause the soil or tundra or oceans to release carbon dioxide and methane, both potent greenhouse gases. This set of feedbacks, which will ultimately shape much of our fate, is discussed in detail in chapter 3.

Given all these vicious cycles, you would expect to see the temperature record of the past several hundred thousand years march in lockstep with the level of greenhouse gases such as carbon dioxide (CO_2) and methane (CH_4). And as figure 2 illustrates, this is precisely what has happened. Here is the historical record of carbon dioxide alongside temperature for the last 400,000 years. The rise in carbon dioxide at the start of every interglacial warming period trails the temperature rise by a few hundred years. The warming appears to be initiated by changes in Earth's orbit around the sun, which in turn leads to increases in carbon dioxide (and methane), which then accelerate the warming, which increases the emissions, which increases the warming. . . .

This is one of the most revealing as well as astonishing graphs ever compiled.

The high confidence that scientists have in these records in-

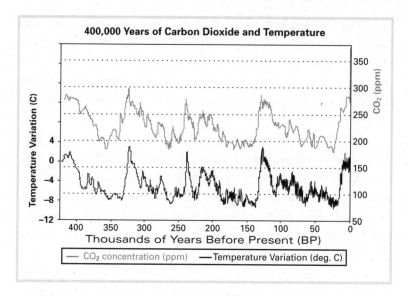

Figure 2. The record of atmospheric concentrations of carbon dioxide, CO_2, in parts per million or ppm (upper curve), together with the temperature record (lower curve), for the past 400,000 years.

The sharp increase in CO_2 concentrations since 1900 is so unprecedented in the historical record, it appears simply as a vertical line extending from 280 to 380 ppm at the far upper-right of the chart. The level of CO_2 is now far outside the bounds humans have ever experienced.

The close coupling between CO_2 and temperature over the past several hundred thousand years is one of many reasons scientists are confident that global temperatures this century will keep rising faster and faster as concentrations of CO_2 rise faster and faster. We have warmed about 0.8°C since the industrial revolution (not shown on this chart).

creased further in 2005 when researchers examined a 2-mile-long Antarctic ice core that extended the records—and extended the tight correlation between temperature and greenhouse gas concentrations—back even farther in time. As NASA's Gavin Schmidt wrote of that research, "That a number of different labs, looking at ice from different locations, extracted with different methods, all give very similar answers is a powerful indication that what they are measuring is real."

The past is prologue. This paleoclimate record is an eye-opener and a heart-stopper. It reveals the enormous risks we are taking with our planetwide, damn-the-consequences climate experiment. Because of human activity, mostly the combustion of fossil fuels, Earth's carbon dioxide levels are literally off the chart. So are methane levels.

Carbon dioxide levels in particular are higher than they have been for millions of years. The rate of increase is even more worrisome than the levels. Levels are rising 200 times faster than at any time in the last 650,000 years. If "Earth's climate system is an ornery beast which overreacts to even small nudges," as Dr. Broecker put it, what will happen to people foolish enough to keep punching it in the face?

On our current emissions path, Earth's average temperature will probably rise 1.5°C by midcentury. By century's end we will be more than 3°C warmer than today. *The last time Earth was 1°C warmer than today, sea levels were 20 feet higher.* That occurred during the Eemian interglacial period about 125,000 years ago, when Greenland appears to have had far less ice.

How fast can the sea level rise? Following the last ice age, the world saw sustained melting that *raised sea levels more than a foot a decade.* Many scientists believe we could see such a melting rate—a catastrophic melting rate of more than 12 inches every ten years—within this century. Sea levels ultimately could rise much more than 20 feet because Antarctica contains far more landlocked ice than Greenland.

The last time Earth was 2° to 3°C warmer than it is now, some 3 million years ago, sea levels were more than 80 feet higher.

THE ANSWER TO THE QUESTION OF THE CENTURY

Carbon dioxide, CO_2, is the principal greenhouse gas forcing the climate to change. In the past 250 years, industrial processes, mainly burning fossil fuels, have released some 1,100 billion tons of CO_2 into the atmosphere cumulatively. Fully half these emissions have occurred only since the mid-1970s, which is why the climate has begun to change so dramatically in recent decades. In 2005, emissions of CO_2 generated by fossil fuel combustion amounted to more than 26 billion tons.

While *emissions* might be thought of as the rate of water flowing into a bathtub, atmospheric *concentrations* are the water level in the bathtub. Concentrations are what affect the climate. Global concentrations of carbon dioxide in the atmosphere had risen slowly from a preindustrial average of about 280 parts per million (ppm) to about 315 ppm by 1960. In 2005, concentrations have soared to 380 ppm, which is not surprising, since emissions have been soaring. Concentrations are now climbing by more than 2 ppm a year.

We have been adding carbon dioxide and other greenhouse gases to the atmosphere at such a fast clip that the planet's warming has not yet caught up to the full forcing of all those heat-trapping gases. So, if we stopped increasing the concentration of greenhouse gases in the atmosphere right now, Earth would still warm up another 0.6°C. Yet, as we've seen, if we warm more than another 1°C, then a 20-plus-foot sea-level rise becomes the likely scenario.

How much do carbon dioxide emissions have to drop to stop increasing concentrations? For the last few decades nearly 60 percent of the carbon dioxide that we have been adding to the atmosphere has stayed there. The other 40 percent is being taken up by the ocean, vegetation, and soils. To stop concentrations from rising, we have to reduce emissions by more than 60 percent (probably closer to 80 percent) from recent levels. Yet far from dropping, carbon dioxide emissions have instead been rising 2 percent per year for the past decade, thanks to steady population and economic growth combined with an absence of collective action to achieve that growth in an environmentally sustainable manner.

Worse still, China, the world's second-largest (and fastest-growing) emitter, is building coal plants and increasing oil use at an accelerating rate. Worst of all, the world's largest emitter, the United States, has not only refused to reduce its emissions, the Bush administration is actually committed to increasing our emissions and blocking other countries from taking action to reduce theirs (see chapter 5). The U.S. Energy Information Administration projects global emissions will rise more than 50 percent between 2000 and 2030.

By about 2015, the planet will be fully committed to another 1°C warming, even if we could cut emissions 80 percent in the span of a few years after 2015. But of course emissions can't drop that quickly in the real world. Replacing the economy's fossil fuel–based industry entirely might take a century. Rather than focusing on impractical impossibilities, let me focus on political improbabilities, since occasionally political realities can change fast.

Instead of replacing the world's existing fossil fuel–based industry, let's ask what it would take just to replace the projected *growth* in emissions for the next fifty years—growth that is now expected to come primarily from more than a thousand *new* large coal plants and more than a billion *new* cars. How that might be done was spelled out in a 2004 *Science* magazine article by Princeton University researchers Stephen Pacala and Robert Socolow, which I here update and modify.

Imagine if the next president, in concert with the U.S. Congress and all the major nations of the world, developed and developing, embarked on an aggressive *five-decade-long* effort to deploy the best existing and emerging energy technology. Imagine that from 2010 through 2060 the world achieves the following astonishing changes:

1. We replicate, nationally and globally, California's performance-based efficiency programs and codes for homes and

commercial buildings. From 1976 to 2005, electricity consumption per capita stayed flat in California, while it grew 60 percent in the rest of the nation.

2. We greatly increase the efficiency of industry and power generation—and more than double the use of cogeneration (combined heat and power). The energy now lost as waste heat from U.S. power generation exceeds the energy used by Japan for all purposes.

3. We build 1 million large wind turbines (fifty times the current capacity) or the equivalent in other renewables, such as solar power.

4. We capture the carbon dioxide associated with 800 proposed large coal plants (four-fifths of all coal plants in the year 2000) and permanently store that CO_2 underground. This is a flow of CO_2 *into* the ground equal to the current flow of oil *out* of the ground.

5. We build 700 large nuclear power plants (double the current capacity) while maintaining the use of all existing nuclear plants.

6. As the number of cars and light trucks on the road more than triples to 2 billion, we increase their average fuel economy to 60 miles per gallon (triple the current U.S. average) with no increase in miles traveled per car.

7. We give these 2 billion cars advanced hybrid vehicle technology capable of running on electricity for short distances before they revert to running on biofuels. We take one-twelfth of the world's cropland and use it to grow high-yield energy crops for biofuels. We build another half-million large wind turbines dedicated to providing the electricity for these advanced hybrids.

8. We stop all tropical deforestation, while doubling the rate of new tree planting.

Other strategies exist, but I consider them more challenging and improbable than any of these.

If we succeeded in every single one of these eight monumental efforts, keeping global CO_2 emissions frozen at 2010 levels for fifty years, and then we somehow were able to sharply *decrease* global emissions starting in 2061, we would stabilize concentrations at about 550 ppm. In this scenario, temperatures would still rise steadily over the course of the century by an additional 1.5°C or more, with further warming after 2100. The Greenland Ice Sheet would likely still melt, with the resulting 20 feet of sea-level rise— but we would have slowed the process significantly and perhaps avoided the worst of the sea-level rise, 40 to 80 feet or more (assuming that we have also adopted strong policies to constrain the emissions of methane and all other greenhouse gases).

This strategy saves the world from misspending trillions of dollars in polluting, inefficient capital over the next quarter-century (in traditional coal plants, gas-guzzling vehicles, and the like). Most important, *it buys the world time to achieve an even stronger consensus for action,* which in turn could lead to a far more accelerated rate of technology deployment after, say, 2030. And that could potentially keep concentrations below 500 ppm and save most of the Greenland Ice Sheet.

Obviously, and tragically, the chances are slim that we will start pursuing these eight changes in 2010. Right now, we don't have the political consensus in this country to begin pursuing even one of them. You may think some of them are implausible, yet none is technically impossible right now. This strategy is the best way to avoid Hell and High Water while expanding living standards at home and around the world—and, as we will see, it is far easier than the alternative strategy we face if we delay much longer.

Pacala and Socolow published their study to show that "humanity already possesses the fundamental scientific, technical, and industrial know-how to solve the carbon and climate problem for the next half-century." The tragedy, then, as historians of the future will most certainly recount, is that we ruined their world not because we lacked the knowledge or the technology to save it but *simply because we chose not to make the effort.*

This scenario might be called "Two Political Miracles" because it would require a radical conversion of American conservative leaders—first, to completely accept climate science, and second, to strongly embrace a variety of climate solutions, most of which they currently view as anathema. To repeat, we lack not the technology but the political will.

The answer to the question of the century—Do we humans have the political will to stop the great ice sheets from melting?—is, at best, "Not yet."

A NOTE ON TEMPERATURES

Reported temperature changes from global warming can be a source of some confusion. Americans use the Fahrenheit scale of temperature and have the most intuitive familiarity with it. I will also use the Centigrade scale because most scientific research uses it. Anyone who wants to become knowledgeable about global-warming research needs to become familiar with thinking in Centigrade terms.

Since I'm focusing on temperature change, here is the key conversion: A 1°C change equals a 1.8°F change. Thus a 5°C change equals a 9°F change—not quite double.

Different parts of the globe are expected to warm up at faster or slower rates than the global average. The land typically warms up faster than the oceans, and higher latitudes warm up faster than the tropics. Most of the inland continental United States is expected to warm up roughly 50 percent faster than the global average. So an additional average global warming of

2°C (3.6°F) means much of this country would be expected to warm 3°C (4.8°F).

Confusion can arise when scientists report how much warming will result from a rise in greenhouse gas concentrations. Some report how much the temperature will rise *from preindustrial temperature levels,* while others talk about how much additional or further warming will occur *from present-day levels.* We have already warmed 0.8°C through 2005, so the difference is significant. I will state which measure I am using each time.

Most analyses suggest that a doubling of greenhouse gas concentrations from preindustrial levels will increase global temperatures about 3°C from preindustrial levels, which is 2.2°C warming from current levels (although many studies suggest the climate could be even more sensitive to a doubling of CO_2 concentrations, as we will see).

A related confusion: Some scientists report how much the temperature will rise ultimately (due to a given rise in greenhouse gas concentrations), while others report only how much the temperature will rise by 2100. Because of the lags in the climate system, those figures can be quite different. I will usually describe how much temperature will rise by 2100.

CHAPTER TWO

2000–2025: REAP THE WHIRLWIND

I don't see any reason why the power of hurricanes wouldn't continue to increase over the next 100 to 200 years.

> —Kerry Emanuel, MIT atmospheric
> scientist, 2006

On our current warming trend, four super-hurricanes—category 4 or stronger—a year in the North Atlantic is likely to become the norm 20 years from now.

> —Judith Curry, Georgia Tech atmospheric
> scientist, 2006

On August 23, 2005, a tropical depression formed 175 miles southeast of Nassau. By the next day, it had grown into Tropical Storm Katrina and was intensifying rapidly. Early in the evening on August 25, Hurricane Katrina made landfall near North Miami Beach. Even though it was only a category 1 storm, with sustained wind speeds of about 80 mph, it caused significant damage and flooding and took fourteen lives.

The hurricane's quick nighttime trip across Florida barely fazed the storm. Entering the Gulf of Mexico's warm waters quickly kicked Katrina into overdrive, like a supercharged engine on high-octane fuel. Hurricanes fuel themselves by continually sucking in and spinning up warm, moist air.

On August 28, Katrina reached category 5 status, with sustained wind speeds of 160 mph and a pressure of 908 millibars. A few hours later, wind speeds hit 175 mph, which they maintained until the afternoon.

At 4 p.m. the National Hurricane Center warned that local storm surges could hit 28 feet, and "some levees in the Greater New Orleans Area could be overtopped," a warning that was tragically ignored by federal, state, and local emergency officials. Over the next fourteen hours, Katrina's strength dropped steadily. When the hurricane's center made landfall Monday morning, it was a strong category 3, battering coastal Louisiana with wind speeds of about 127 mph. The central pressure of 920 millibars was the third-lowest pressure ever recorded for a storm hitting the U.S. mainland.

The devastation to the Gulf region was biblical. The death toll exceeded 1,300. The damage exceeded $100 billion. A half-million people were forced to leave their homes, *more than were displaced during the 1930s dust bowl migration.* One of the nation's great cities was devastated.

About 20 miles to the east of the second Gulf landfall is the small town named Pass Christian, Mississippi, where my brother lived with his wife and son. Tropical cyclones in the Northern Hemisphere rotate counterclockwise, and so the most intense storm surge is just to the east of the eye, because the surge represents the intense winds pushing the sea against the shore. A 30-foot wall of water with waves up to 55 feet high crashed over the town. Although my brother and his family lived a mile inland, their house was ravaged by water up to 22 feet high, leaving its contents looking like they had been churned "inside of a washing machine," in my brother's words. While they lost virtually all their possessions, they were safe in a Biloxi shelter.

Thanks to the generosity of many people, my brother's family was able to find a temporary home in Atlanta. But like many families whose lives were ripped apart by the storm, they had difficult choices in the ensuing months. Perhaps the toughest decision was whether to rebuild their home or to uproot themselves and try to create a new life somewhere else.

I very much wanted to give my brother an expert opinion on what was likely to come in the future. After all, climate change was my field, and while my focus has been on climate solutions, I had done my Ph.D. thesis on physical oceanography.

As I listened and talked to many of the top climate experts, it quickly became clear that the climate situation was far more dire than most people—and even many scientists, myself included— realized. Almost every major climate impact was occurring faster than the computer models had suggested. Arctic sea ice was shrinking far faster than every single model had projected. And the great ice sheets of Greenland and West Antarctica were shedding ice decades earlier than the models said. Ecosystems appeared to be losing their ability to take up carbon dioxide faster than expected. At the same time, global carbon dioxide emissions and concentrations were rising faster than most had expected.

As for hurricanes, global warming had been widely projected to make them more intense and destructive, but again the recent increase in intensity was coming sooner than the computer models had suggested. Why is that a concern? Since 1970, the temperature of the Atlantic Ocean's hurricane-forming region has risen 0.5°C (0.9°F). Over the path of a typical hurricane, this recent ocean warming added the energy equivalent of a few hundred thousand Hiroshima nuclear bombs. On our current emissions path, the Atlantic will warm *twice as much,* another 1°C, by midcentury, and perhaps another 2°C beyond that by century's end. Who can even imagine the hurricane seasons such warming might bring?

This is what I ultimately told my brother, the same advice I would give anyone contemplating living near the Gulf Coast:

Only a quarter of Atlantic hurricanes make U.S. landfall, and while there is no question that the frequency of intense Atlantic hurricanes is rising, where they will actually go any given year is somewhat random.

That said, the Gulf of Mexico is going to get warmer and warmer, as is the Atlantic Ocean, and so hurricanes that enter the Gulf are likely to start out and end up far more destructive than usual. I would not bet that the Mississippi Gulf Coast will get hit by a super-hurricane in any particular year, but I would certainly plan on it being hit again sometime over the next ten years; I wouldn't be surprised if it were hit by more than one.

Coastal dwellers from Houston to Miami are now playing Russian roulette with maybe two bullets in the gun chamber each year. In a couple of decades, it may be three bullets.

Some argue that the recent jump in severe hurricanes was caused by a rise in sea-surface temperatures that is just part of a natural cycle. That position is scientifically untenable, which is why most of the people who advance it are not global-warming researchers. We'll see why the natural-cycles argument will no doubt prove to be "largely false," as MIT's Kerry Emanuel said in 2006. Hurricane seasons with four or more super-hurricanes—those with sustained wind speeds of 131 mph or more—will soon become the norm.

THE ERA OF EXTREME WEATHER

Scientists have long known that global warming increases the chances for extreme weather events. Here's how: As it gets hotter, summer heat waves become longer, hotter, and more widespread. Dry areas tend to dry out faster and to stay that way for longer periods. The extra heat puts more water into the atmosphere, and that causes wet areas to become wetter and annual rainfall to become more intense, which, coupled with earlier snowmelt, leads to more flooding. And hurricanes, which feed on warm seas and atmospheric moisture, become more intense.

This well-accepted scientific theory of how global warming should change the weather has begun shifting to grim reality—our weather is changing, and not for the better. In July 2003 the World Meteorological Organization (WMO) cataloged a number of extreme events: Switzerland had experienced the hottest June in "at least the past 250 years," and the United States had suffered 562 tornadoes in May, exceeding the previous record of 399 in June 1992. The WMO linked them to global climate change. As *The Independent* newspaper of London put it, the WMO "signalled last night that the world's weather is going haywire." The WMO, an "organisation that is not given to hyperbole," noted, "New record extreme events occur every year somewhere on the globe, but in recent years the number of such extremes have been increasing." Since that WMO report, Europe has experienced even more extreme events, including an extended heat wave that caused more than 35,000 deaths in August 2003.

In 2005 the weather became even more hellish. The year was the hottest in recorded history, according to NASA's Goddard Institute of Space Studies. In September the Arctic had the smallest amount of sea-ice cover ever recorded by satellites. Mumbai, India, saw that country's most intense recorded instance of rainfall—3 feet of rain in twenty-four hours.

The extremes of wet and dry are astounding. While southern Louisiana was deluged with rain in the summer of 2005, a record-smashing U.S. hurricane season, "the eight months since October 1, 2005," were its driest "in 111 years of record-keeping," the National Climatic Data Center reported in July 2006. While in 2005 much of the Northeast drowned in the wettest October in recorded history, the United States as a whole had its worst wildfire season.

You may reasonably ask, Don't extreme conditions happen somewhere on the planet all the time? How do we know this weather is truly out of the ordinary?

As far back as 1995, analysis by the National Climatic Data Center showed that over the course of the twentieth century, the United States had suffered a statistically significant increase in a variety of extreme weather events, the very ones you would expect from global warming, such as more—and more intense—precipitation. That analysis also concluded that the chances were only "5 to 10 percent" that this increase was due to factors *other* than global warming, such as "natural climate variability." And since 1995 the climate has gotten *much* more extreme.

A 2004 analysis by the center found an increase during the twentieth century of "precipitation, temperature, streamflow, heavy and very heavy precipitation and high streamflow in the East." It found a 14 percent increase in "heavy rain events" of more than 2 inches in one day, and a 20 percent increase in "very heavy rain events"—best described as deluges—more than 4 inches in one day. These extreme downpours are precisely what is predicted by global-warming scientists and models. The deluge that socked the Mid-Atlantic states and the Northeast the last week of June 2006 fits the picture of this global-warming-type rainstorm. Washington, D.C., for instance, was drenched by more than 7 inches of rain in one twenty-four-hour period. And this deluge happened at the same time that 45 percent of the continental United States was experiencing moderate to extreme drought, which is far above the historical norm.

The center, a division of the National Oceanic and Atmospheric Administration (NOAA), which is part of the U.S. Department of Commerce, has developed the U.S. Climate Extremes Index to quantify these climate changes. The index measures the percentage of the country that is subject to a variety of extreme conditions, including:

- much higher (and lower) than normal maximum temperatures
- much higher (and lower) than normal minimum temperatures
- severe drought and severe excess moisture
- an extreme proportion of total rain from intense one-day rainstorms
- much greater than normal number of days of the year with precipitation or without precipitation

It averages each of these five factors with a sixth one—the frequency and intensity of tropical storms making U.S. landfall. The index uses a scale from 0 to 100; 100 means the whole country has extreme conditions throughout the year *for each of the indicators,* "a virtually impossible scenario," the center notes.

The index extends from 1910 to today, during which time the average has been 20. The most extreme year was 1998, with an index of nearly 44, more than double the average. The second-most extreme year was 2005, with an index of about 41. The seventeen least-extreme years of the past century all came before 1980.

The index almost certainly *underestimates* how much the country is suffering the impact of global warming—for two reasons. First, it averages in some extreme conditions that are occurring *less* often because of warming, such as the "percentage of the United States with minimum temperatures much *below* normal." Second, the index excludes Alaska—the largest state and the one suffering the most extreme climate change. For instance, a 2003 report by the General Accounting Office found that "flooding and erosion affects 184 out of 213, or 86 percent, of Alaska Native villages . . . due in part to rising temperatures." *Half or more of the villages may need to be relocated.* In January 2005 the city of Valdez, Alaska, hit 54°F, beating the city's previously warmest January day by 8°.

If the weather is becoming more extreme, what is happening to the most extreme weather events, like hurricanes? The scientists and studies I find most credible conclude that "greenhouse warming is causing an increase in global hurricane intensity." To explain why this is almost certainly true, I will rely on the recent work of scientists at MIT, the Georgia Institute of Technology, the National Center for Atmospheric Research, and NASA's Goddard Institute of Space Studies, and my conversations with many of those scientists. The framework of this explanation comes from a 2006 paper written by Judith Curry and others. The central hypothesis is best explained by dividing it into a causal chain of three sub-hypotheses.

1. GLOBAL TROPICAL SEA-SURFACE TEMPERATURE IS INCREASING AS A RESULT OF HUMAN-CAUSED GREENHOUSE WARMING

A January 2006 report on Katrina by the National Climatic Data Center noted, "There has been an overall increasing trend in July–September Atlantic and Gulf of Mexico sea surface temperatures during the past 100 years. . . . This pattern is similar to that observed across global land and ocean surfaces."

The planet is warming—especially the oceans. Since 1955 the oceans have absorbed roughly *twenty times* more heat than the atmosphere. A team of scientists led by NASA's James Hansen have actually measured the increasing ocean heat content over the past decade. They reported in 2005 that it matches the predicted warming from greenhouse gases.

Another 2005 study, this one led by the Scripps Institution of Oceanography, compared actual ocean-temperature data from the surface down to hundreds of meters (in the Atlantic, Pacific, and Indian Oceans) with climate models and concluded:

A warming signal has penetrated into the world's oceans over the past 40 years. The signal is complex, with a vertical structure that varies widely by ocean; it cannot be explained by natural internal climate variability or solar and volcanic forcing, but is well simulated by two anthropogenically forced climate models. We conclude that it is of human origin, a conclusion robust to observational sampling and model differences.

Anthropogenic is science-speak for "caused by humans." Greenhouse gases, such as carbon dioxide from fossil fuel combustion, are forcing the earth's climate to warm. Even at a depth of 600 feet, the North Atlantic has warmed 0.2°C thanks to human emissions.

The science gets stronger every year. A comprehensive 2006 analysis using the climate model of the Goddard Institute for Space Studies found that greenhouse gas forcings explain nearly four-fifths of the warming in the main region where Atlantic tropical storms start, and human-generated emissions also account for all the warming in the Gulf of Mexico. Other recent scientific analyses come to similar conclusions.

This gives us very solid science for concluding that *global tropical sea-surface temperature is increasing as a result of human-caused greenhouse warming.*

To disprove this statement, a scientist must not merely come up with an alternative explanation for the remarkable recent warming but be able to identify some as yet unknown and unmeasured effect that is simultaneously negating the well-understood warming from greenhouse gases. Nobody has yet done either.

2. AVERAGE HURRICANE INTENSITY INCREASES WITH INCREASING TROPICAL SEA-SURFACE TEMPERATURE

Of the three, this statement is the most scientifically straightforward. As Kerry Emanuel wrote in his 2005 book, *Divine Wind: The History and Science of Hurricanes:* "By trapping heat energy in the ocean, the greenhouse effect sets the stage for the meteorological explosion that is the hurricane." Many factors must coexist to create hurricanes, which are like sophisticated race cars, but a hurricane's engine can't start without warm water to give it a steady supply of fuel.

More than fifty years ago, scientists established that tropical cyclones form only if sea-surface temperatures (SST) exceed 80°F. Absent the natural greenhouse effect, which keeps the planet 60°F warmer than it otherwise would be, we would not have hurricanes. Both theory and observation, including several recent studies, support the relationship between sea-surface temperature and hurricane intensity.

How did Katrina turn into a powerful category 5 hurricane? The National Climatic Data Center 2006 report on Katrina begins its explanation by noting that SSTs in the Gulf of Mexico during the last week in August 2005 "were one to two degrees Celsius above normal, and the warm temperatures extended to a considerable depth through the upper ocean layer." The report continues, "Also, Katrina crossed the 'loop current' (belt of even warmer water), during which time explosive intensification occurred. The temperature of the ocean surface is a critical element in the formation and strength of hurricanes."

An important factor was that the ocean warming had penetrated to a considerable depth. One of the ways that hurricanes are weakened is the upwelling of colder, deeper water due to the hurricane's own violent action. But if the deeper water is also warm, it doesn't weaken the hurricane. In fact, it may continue to intensify. Global warming heats both the sea surface and the deep water, thus creating ideal conditions for a hurricane to survive and thrive in its long journey from tropical depression to category 4 or 5 superstorm.

After Katrina, Georgia Tech scientists reexamined the historical hurricane and SST data using "a methodology based on information theory, isolating the trend from the shorter term natural modes of variability." They looked at four factors that can affect hurricane intensity: atmospheric humidity, wind shear (which can rip storms apart), rising SSTs, and large-scale air-circulation patterns. "Results show that the increasing trend in the number of category 4 and 5 hurricanes for the period 1970–2004 is directly linked to the trend in SSTs; other aspects of the tropical environment, while influencing shorter term variations in hurricane intensity, do not contribute substantially to the observed global trend."

The evidence gives us a high level of confidence that statement number 2 is true: Average hurricane intensity increases with increasing tropical sea-surface temperature.

3. THE FREQUENCY OF THE MOST INTENSE HURRICANES IS INCREASING GLOBALLY

Three major articles published in mid-2005 pointed out that intense hurricanes had become more common in recent decades. These analyses spawned a whirlwind of media attention because the authors were highly credible and because the articles happened to come out in the weeks before and just after Katrina.

Dr. Kevin Trenberth, head of the Climate Analysis Section of the National Center for Atmospheric Research (NCAR), published the first, in *Science,* two months before Katrina. He began by noting that in 2004 "an unprecedented four hurricanes hit Florida; during the same season in the Pacific, 10 tropical cyclones or typhoons hit Japan (the previous record was six)." What we call hurricanes in America are called cyclones or typhoons in other parts of the world. They all have maximum sustained surface winds of at least 74 mph. Trenberth explained that theory suggests global warming will increase the intensity of hurricanes and the rainfall they bring. He noted that from 1995 to 2004, Atlantic hurricane seasons were abnormally active, as measured by the Accumulated Cyclone Energy Index, which tracks "the collective intensity and duration of tropical storms and hurricanes" during each season.

Kerry Emanuel, a professor of atmospheric sciences at MIT, published next, in *Nature,* a few weeks before Katrina hit. Emanuel, one of the world's leading hurricane experts, created a measure of hurricane destructiveness, which he called the *power dissipation index.* This is essentially the maximum sustained wind speed cubed (raised to the third power)—a measure of hurricane intensity that correlates well with the "actual monetary loss in windstorms"—integrated over the storm's life. He then used the best available data from all sources for hurricanes and sea-surface temperature in both the North Atlantic and the western North Pacific.

Emanuel found a sharp increase in the index in the last thirty years and a close correlation between the power dissipation index and SST in both oceans. Tropical cyclones in both oceans have increased both their peak wind speed and their duration substantially since 1949.

Finally, in September, scientists from Georgia Tech (Peter Webster, Hai-Ru Chang, and Judith Curry) and NCAR (Greg Holland) published in *Science* a detailed analysis of hurricanes in six different ocean regions, including the North Atlantic. They examined the record for the past thirty-five years, the period when high-quality satellite data became available. During this time, SSTs increased about 0.5°C. They found a large increase in the number of super-hurricanes (categories 4 and 5) in every region. Comparing the

1975–1989 period with the 1990–2004 period, they found a more than 50 percent increase in super-hurricanes overall and in the North Atlantic. They concluded that "global data indicate a 30-year trend toward more frequent and intense hurricanes, corroborated by the results of the recent regional assessment [Emanuel's 2005 study]."

THE FUTURE IS NOW

The terms *hypothesis* and *theory* are often used interchangeably, but for scientists, a theory is "a hypothesis that has been confirmed or established by observation or experiment, and is propounded or accepted as accounting for the known facts; a statement of the general laws, principles, or causes of something known or observed," as the *Oxford English Dictionary* defines it. Theories have heft. They have credibility. The germ-theory of disease and human-caused global warming are well-established scientific theories.

For a hypothesis like "Greenhouse warming is causing an increase in global hurricane intensity" to be elevated to theory status, it must pass three additional tests, beyond accounting for the observed data. A theory must make accurate predictions, survive scrutiny by critics, and beat out competing theories, as Judith Curry has written. Let's consider the theory's predictive value.

The three papers described above, arguing that an increase in SST was causing an increase in intense hurricanes, were all based on data through 2004. Since 2005 turned out to be the warmest year on record, with high June–November SSTs in the Atlantic and Gulf of Mexico, it is valuable to examine some of the remarkable records set that year, courtesy of the National Oceanic and Atmospheric Administration:

- Twenty-seven named tropical storms—from Arlene to Wilma, Alpha to Zeta—formed during the 2005 season. This is the *most* named storms in a single season, breaking the old record of 21 set in 1933.

- Fifteen hurricanes formed during the 2005 season (a post-storm analysis in 2006 upgraded Cindy from a tropical storm to a hurricane). This is the *most* hurricanes in a single season, breaking the old record of 12 set in 1969.
- Seven category 3 or higher hurricanes formed during the 2005 season. This *ties* the season record for such hurricanes, first set in 1950.
- Four category 5 hurricanes formed during the 2005 season (Emily, Katrina, Rita, and Wilma). This is the *most* category 5 hurricanes recorded in a single season, breaking the old record of 2 set in 1960 and 1961.
- Seven named storms made United States landfall during 2005. This puts the 2005 season in a *tie* for second place for landfalling storms, behind the 1916 and 2004 seasons where eight named storms made landfall. An eighth storm brushed the coast of North Carolina in 2005 but did not make an official landfall.
- The 2005 season was the *most* destructive for United States landfalling storms, largely due to Katrina. Damage estimates for the 2005 season are over $100 billion.
- Dennis became the *most* intense hurricane on record before August when a central pressure of 930 mb was recorded.
- Emily *eclipsed* Dennis's record for lowest pressure recorded for a hurricane before August when its central pressure dropped to 929 mb. Emily's strength was revised in 2006, so it became "*the earliest-forming Category 5 hurricane on record* in the Atlantic basin and the only known hurricane of that strength to occur during the month of July."
- Vince was the *first* tropical cyclone in recorded history to strike the Iberian Peninsula. Vince was the farthest north and east a storm has ever developed in the Atlantic basin.

In the end, 2005 was not just the warmest year on record, it had the most intense and long-lasting hurricane season, as measured by the Accumulated Cyclone Energy Index. One hurricane season cannot, however, confirm or disprove this hypothesis (or competing hypotheses). Hurricane seasons are subject to enormous year-to-year variability because of factors such as the El Niño weather pattern, which tends to weaken Atlantic hurricane seasons. But we should expect a general upward trend in the intensity and length of Atlantic hurricane seasons, and we should expect more and more records to be smashed.

A strong hypothesis is hard to criticize effectively and objectively; a weak one is not. Let's see how this one fares. The critique offered by meteorologists in particular is worth exploring in detail because it sheds light on the national global-warming debate and on how the nation is likely to respond to the growing evidence of climate change over the next decade or two.

The first major critique of the theory and the 2005 studies supporting it was "Hurricanes and Global Warming," published in the November 2005 issue of the *Bulletin of the American Meteorological Society*. Among its coauthors were three leading public critics of the warming-hurricane connection—Max Mayfield, director of NOAA's National Hurricane Center, Christopher W. Landsea of NOAA's Hurricane Research Division, and Roger Pielke Jr. of the University of Colorado at Boulder—together with two other NOAA hurricane experts. A subhead that begins the article, "An interdisciplinary team of researchers survey the peer-reviewed literature to assess the relationships between global warming, hurricanes, and hurricane impacts," is followed by:

> Debate over climate change frequently conflates issues of science and politics. Because of their significant and visceral impacts, discussion of extreme events is a frequent locus of such

conflation. Linda Mearns, of the National Center for Atmospheric Research (NCAR), aptly characterizes this context: "There's a push on climatologists to say something about extremes, because they are so important. But that can be very dangerous if we really don't know the answer."

Wow! I have read hundreds of literature-survey articles by scientists over the years, and not a single one began like that. You don't have to be a scientist to realize that objective surveys don't start by questioning the character of those they disagree with—their motives and their scientific method. These authors suggest that Emanuel and the others have a political agenda, and rather than presenting sound analysis, they have been pushed to say things they can't support.

You would never know from this article (four of whose authors work at NOAA) that a division of NOAA, the National Climatic Data Center, had been repeatedly publishing articles and an index showing that extreme events are in fact becoming more frequent. You would also never know from this article that there is a strong consensus among climate scientists that global warming leads to more extreme weather events. The authors never report that information or even discuss the subject. They just imply that those who make such arguments are not practicing pure science.

While claiming to be an up-to-date survey, the article bases most of its critique on old studies that largely predate the recent surge in SSTs and hurricane intensity examined in the new studies. When it does examine the new studies, it focuses primarily on Atlantic hurricanes, even though the second *Science* paper found that "the largest increase [in category 4 and 5 hurricanes] occurred in the North Pacific, Indian, and Southwest Pacific Oceans." There is no explanation for this omission. For the specific matter of the North Atlantic, the authors assert that "much of the recent upward trend in Atlantic storm frequency and intensity can be attributed to large multidecadal fluctuations," although the authors never define "much."

This brings us to the competing hypothesis: Atlantic SSTs and hurricanes come in natural multidecadal cycles. This hypothesis deserves attention because the natural-cycles argument is repeated widely in the media, with arguments such as "We had some big hurricanes in the 1940s" used to imply that what we are seeing today is not evidence of human-caused climate change.

NATURAL CYCLES VERSUS GLOBAL WARMING

We have two battling hypotheses. In one corner is the global-warming theory, which says that forcings (natural and human-made) explain most of the changes in our climate and temperature. The natural forcings include fluctuations in the intensity of sunlight (which can increase or decrease warming) and major volcanoes that inject huge volumes of gases and aerosol particles into the stratosphere (which tend to block sunlight and cause cooling). The biggest forcings caused by humans are the greenhouse gases we generate, particularly carbon dioxide from burning coal, oil, and natural gas. But we humans also put significant sulfate aerosols into the atmosphere from burning coal and diesel fuel without advanced emissions controls.

Global warming explains the vast majority of the recent warming in North Atlantic SSTs and most, if not all, of the rise in hurricane intensity in all oceans. A 2006 article in the *Bulletin of the American Meteorological Society* by six leading climatologists noted that recent research "specifically shows an increase [in hurricane intensity] in all ocean basins and an overall global increase, which is the type of signature that would be expected from global warming changes."

In the other corner, we have the natural-cycles hypothesis. This hypothesis offers little or no explanation for the rising hurricane activity in the North Pacific, Indian, and southwestern Pacific Oceans. Nor can it account for most of the rise in North Atlantic SSTs over the past three decades. Its advocates claim it can explain much of the high level of Atlantic hurricane activity in the 1940s, 1950s, and early 1960s and now again in the past decade.

The natural cycle in the Atlantic is called the Atlantic multi-decadal oscillation. Consider figure 3, which is a plot of average sea-surface temperatures for June–November in the North Atlantic's hurricane-forming region. You can just make out what looks like a 60- to 70-year cycle with positive peaks around the 1880s and 1950 (and possibly 2005) and negative peaks around the early 1910s and the mid-1970s. As you can see, however, the dip around the early 1910s is much deeper than the little dip centered in the mid-1970s. Similarly, the peak of each cycle keeps getting higher. Not surprisingly, the 1995–2005 period has had considerably more total tropical storms, hurricanes, and category 4 and 5 hurricanes—the city destroyers—than the peak years from the last cycle of Atlantic hurricane activity (1945–1955).

When I first began researching hurricanes, I believed, like many scientists, that global warming made hurricanes more intense *and* that hurricanes followed a natural oscillation six to seven decades long. I knew that climate scientists had an excellent understanding of the shape of the entire temperature record, including the temperature peak in midcentury, as detailed in several recent modeling studies of the various natural and human-made forcings. But I never connected the dots.

Then, at an October 2005 seminar of the American Meteorological Society, MIT's Kerry Emanuel pointed out that the downswing in temperature and hurricane activity in the 1960s and 1970s may not have been "a natural fluctuation," adding that "a lot of what I thought was natural I now think was forced." Also, Judith Curry from Georgia Tech presented slides showing a stunning parallel be-

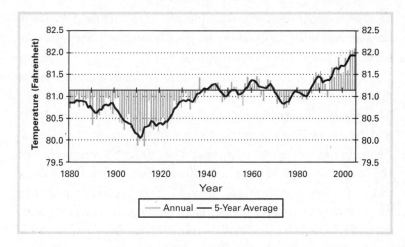

Figure 3. The average sea-surface temperature in °F for June–November in the North Atlantic's hurricane-forming region, since 1880. A 5-year mean is also provided to smooth out annual fluctuations. The data is from NOAA.

tween SSTs and hurricane intensity in the Atlantic over the last century.

As I talked with more climate experts and reviewed the literature, I saw that the dots—the multiple forcings of the past hundred years together with the temperature and hurricane trends—formed a clear picture. Why the temperature dip centered in the early 1910s? A series of six major volcanoes erupted from 1875 to 1912 all around the globe: in Iceland in 1875, Indonesia (Krakatoa) in 1883 (the largest explosion ever recorded), New Zealand in 1886, Guatemala in 1902, Kamchatka in 1907, and Alaska in 1912. The aerosols emitted by these awesome volcanoes kept the planet cooler than it would have been during this time.

The subsequent rise in global temperatures and Atlantic sea-surface temperatures is also well explained by forcings—a slow but steady increase in human-generated greenhouse gases, a slight increase in solar intensity, and the absence of any major volcano eruptions that might otherwise have blocked these trends. The rise in number and intensity of hurricanes in midcentury occurred at the same time as this rise in Atlantic SSTs.

But what explains the drop in SSTs and hurricanes from the mid-1960s through the early 1990s that some critics (mistakenly) believe undercuts the global-warming theory? Not coincidentally, that drop began just around the time of the 1963 eruption of Mount Agung in Indonesia, which "produced the largest stratospheric dust veil in the Northern Hemisphere in more than 50 years." Not coincidentally, the drop also came during a three-decade stretch when humans were emitting unprecedented amounts of industrial aerosol and sulfate. Not coincidentally, the drop continued through the 1982 El Chichon eruption in Mexico and the 1991 Mount Pinatubo eruption in the Philippines, which "produced very large stratospheric aerosol clouds and large climatic effects." Multiple major sun-blocking events all worked together to give us a false sense of security, to shield us from the full impact of the rapid growth in atmospheric carbon dioxide concentrations. And they cooled the air and the seas, resulting in fewer intense hurricanes.

Not coincidentally, temperatures (and hurricanes) rebounded strongly after the major volcanic eruptions ended and the human-made aerosol emissions by the industrial nations dropped sharply thanks to clean-air regulations, even as human emissions of greenhouse gases continued to soar and utterly overwhelmed the aerosol cooling effect.

With all these variables, no wonder this picture took climate scientists so long to bring into focus. This complexity helps explain why many meteorologists, most of whom have little training in global-warming science, keep standing by their flawed natural-cycles hypothesis.

The view that this is not all a grand coincidence gains credence from a 2006 study by American and British climatologists and meteorologists led by the Lawrence Livermore National Laboratory.

They concluded: "Volcanically induced cooling of the ocean surface penetrated into deeper layers, where it persisted for decades after the event." The research makes clear that the combination of natural and human-made forcings, correctly modeled, can explain the key trends in both SSTs and subsurface ocean temperatures for the past 120 years. Subsurface temperatures can be as important to hurricane intensity as sea-surface temperatures.

When I posed all this to MIT's Kerry Emanuel in February 2006, he replied, "It would appear that Atlantic hurricanes are a kind of global thermometer, following closely the trend in global (and especially Northern Hemispheric) temperatures produced by volcanic activity, solar variations, sulfate aerosols, and greenhouse gases. *I think the 'natural cycles' argument will prove to be largely false.*"

In 2006, Emanuel coauthored a study that concluded that the Atlantic multidecadal oscillation (AMO) trend has been overestimated and that "there is no evidence that natural climate oscillations such as the AMO contributed to long-term tropical North Atlantic SST variations." Global warming is now clearly the dominant force behind SSTs in the Atlantic's hurricane-forming region and will become more and more dominant in the future.

Higher SSTs have helped cause the rapid intensification of hurricanes like Katrina, as the post-Katrina report by the National Climatic Data Center explained. And the day before Katrina struck, the *New Orleans Times-Picayune* reported, "The northern Gulf of Mexico is unusually warm," likely "the result of *relentless high temperatures in recent weeks* along Louisiana's Gulf Coast," according to National Hurricane Center meteorologist Eric Blake.

Global warming also increases the incidence of such heat waves and makes them more intense and long-lasting. And it puts into the atmosphere more warm, moist air—the stuff hurricanes are made of. Shortly before Katrina, NOAA's Christopher Landsea said, "The warmer the sea-surface temperature and the more warm, moist air that is available, the stronger a hurricane can become." After Katrina, Max Mayfield, then director of the Tropical Prediction Center at the National Hurricane Center, told CBS's Bob Schieffer, "We think the best correlation [with hurricane activity] we have here is with the sea-surface temperatures." Yet Mayfield testified to Congress in September 2005, "The increased activity since 1995 is due to natural fluctuations/cycles of hurricane activity, driven by the Atlantic Ocean itself along with the atmosphere above it and not enhanced substantially by global warming."

This conclusion is "untenable," said a major 2006 study by climate scientists from NASA, Columbia University, Yale University, MIT, Lawrence Berkeley National Laboratory, and Argonne National Laboratory. They concluded that "to the degree that hurricane intensification of the past decade is a product of increasing SSTs in the Atlantic Ocean and the Gulf of Mexico, human-made greenhouse gases probably are a substantial contributor."

Yet untenable critiques from meteorologists can seem very credible to the public, as can critiques from those with little training in or knowledge of global-warming research. So it is no wonder we do not seem close to achieving the consensus needed to avert catastrophic climate change. Many major media outlets, including CNN, *USA Today*, and the *Chicago Tribune*, bought this story line—proving you can spin hurricanes backward—and the combined mantras of "It's a natural cycle" and "Those who say otherwise have a political agenda" may keep the public, the media, and policy makers confused for years to come.

Even though only a small fraction of hurricanes make landfall, the global-warming signal is starting to show up. Emanuel notes that "a trend in landfalling intensity is already apparent" when one looks at hurricanes worldwide. Significantly, the National Climatic Data Center developed a measure for the "strength and frequency" of tropical storms and hurricanes striking this country for its Climate Extremes Index. That index had an average value of 20 over the past ninety-five years. The two highest values this measure has seen were in 2004 and 2005, at 80 and 92 respectively. No year before 1985 exceeded 65.

"More than half the total hurricane damage in the U.S. (normalized for inflation and populations trends) was caused by just five events," explains Emanuel. Storms that are category 4 and 5 at landfall (or just before) are what destroy major cities like New Orleans and Galveston with devastating winds, rains, and storm surges. We have seen a more than 50 percent increase in category 4 and 5 storms both globally and in the Atlantic. Where precisely such storms make landfall is random on a year-to-year basis, but over time, more and more will inevitably strike this country, especially as the Gulf gets warmer. And that is without considering the combined impact of more intense hurricanes and sea-level rise.

Tropical cyclones are threshold events—if SSTs are below 80°F, they do not form. Some analysis even suggests there is an SST "threshold [close to 83°F] necessary for the development of major hurricanes." Global warming may actually cause some hurricanes to develop or intensify that otherwise would not have (by raising SSTs above the threshold at the right place or time).

For now, we can't know with confidence whether global warming has caused a specific hurricane to develop or intensify. But we can know with very high confidence that global warming has increased the intensity and rainfall of recent hurricane seasons.

The destruction of New Orleans by Katrina, particularly the breaching of the levees, might have been avoided if the storm had been a little less severe and generated a little less wind and rain. In the case of Katrina, the added intensity from global warming may have been the straw that broke the camel's back, as NCAR's Kevin Trenberth put it in October 2005, the extra push that brought the poorly built levees down.

2000–2025: REAP THE WHIRLWIND

*For they have sown the wind, and they shall reap the
whirlwind.*

—Hosea

How will the rest of this era play out? Why did I advise my brother
not to rebuild on the Gulf Coast? Why should New Orleans not be
rebuilt unless the levees protecting it are built to withstand a category 5 hurricane?

"I don't see any reason why the power of hurricanes wouldn't
continue to increase over the next 100 to 200 years," said MIT's
Emanuel. Hurricanes can get much, much bigger than we have so
far seen in the Atlantic. The most intense Pacific storm on record
was Super Typhoon Tip in 1979, which reached maximum sustained winds of 190 mph near the center. On its wide rim, gale-force
winds (39 mph) extended over a diameter of an astonishing 1,350
miles. It would have covered nearly half the continental United
States.

No wonder ABC News reported in 2006 that hurricane scientists are considering adding a category 6 for hurricanes above 175
miles per hour. Ultimately, they may become common.

If we don't reverse our emissions paths quickly, global temperatures will rise faster and faster through 2100 and beyond. This will
translate into warmer oceans in all three dimensions: Warmth will
spread over wider swaths of the ocean as well as deeper below the
surface—we've already seen that in the first known tropical cyclone
in the South Atlantic (2004) and the first known tropical cyclone to
strike Spain (2005). That means we will probably see stronger hurricanes farther north along the U.S. Atlantic coast in the coming
decades.

More intense storms will be seen earlier and later in the season. The 2005 hurricane season was the most striking example of that trend, with Emily, "the earliest-forming Category 5 hurricane on record in the Atlantic" in July, and Zeta, the longest-lived tropical cyclone to form in December and cross over into the next year, where it became the longest-lived January tropical cyclone. We have already seen a statistically significant increase in the length of the average hurricane season over the last several decades, according to a 2006 analysis. The data from the past century indicate that a 1°F increase in SSTs leads to an extra five tropical storms a year in the Atlantic—an ominous statistic in a world taking no actions to stop a projected 2°F increase in sea-surface temperatures by midcentury, and more than double that by century's end.

At the same time, the inland United States will heat up at an even faster rate, so the Mississippi River will not be such a cool stream of water pouring into the Gulf. As the sea level rises, the protective outer delta of the Mississippi will continue to disappear and storm surges will penetrate deeper inland. Hurricanes weaken rapidly over land. Even a foot of shallow delta water can dramatically reduce this weakening effect, allowing hurricanes to reach deeper inland with their destructive force.

So not only will we see increased category 4 and category 5 hurricanes, but sooner or later—probably sooner—one of the hurricanes that enters the Gulf will ride a wide and deep mass of warm water straight to the shore, and rather than weakening as it approaches the shore, like Katrina did, it will maintain its strength. Then a category 5 super-hurricane will bring havoc back to New Orleans and the Gulf Coast.

Americans should plan on the 2004 hurricane season—with its four super-hurricanes (category 4 or stronger)—becoming the norm over the next few decades. But if 2004 is the norm, we should not be surprised if as many as a quarter of the hurricane seasons in this era are as severe as those of 2005 with its five super-hurricanes. After all, the ocean and the entire planet are just going to get warmer.

As of the end of 2005, *this decade has already had five of the six hottest years on record* (the other being an El Niño–boosted 1998), so it will no doubt be the hottest decade in thousands of years. As will the next decade. And the decade after that. And on and on and on. Such is the nature of global warming on a planet that refuses to take serious action.

CHAPTER THREE

2025–2050: PLANETARY PURGATORY

Obviously, if you get drought indices like these, there's no adaptation that's possible.
—David Rind, NASA climate scientist, 2005

We're showing warming and earlier springs tying in with large forest fire frequencies. Lots of people think climate change and the ecological responses are 50 to 100 years away. But it's not 50 to 100 years away— it's happening now in forest ecosystems through fire.
—Thomas Swetnam, University of Arizona climate scientist, 2006

Imagine if the climate changed and extreme weather became so constant that it was no longer considered extreme. Mammoth heat waves like the one that killed 35,000 Europeans in 2003 would occur every other year. Mega-droughts and widespread wildfires, like those of the record-breaking 2005 wildfire season, which ravaged 8.5 million acres, would be the norm. This new climate would wipe out whole forests, including virtually every pine tree in British Columbia. The Arctic would have little or no summer ice, and the Greenland ice cap would melt, eventually raising sea levels by 20 feet.

If we permit this Planetary Purgatory to occur, the nation and the world would be forced to begin a desperate race against time—a race against the vicious cycles in which an initial warming causes changes to the climate system that lead to more warming, which makes adapting to climate change a never-ending, ever-changing, expensive, exhausting struggle for our children, and their children, and on and on for generations.

This chapter will focus on (1) the impacts of accelerated warming, especially drought and wildfires, and (2) the fatal feedbacks that will probably start to kick into overdrive during this era and complicate any effort to stop the Greenland Ice Sheet from melting.

HELL AND NO WATER

By the end of the Planetary Purgatory era, 2050, Earth will probably be hotter than it has been in 125,000 years. By then, the planet is likely to be warming 0.6°F (0.33°C) *per decade* or more, even if global-emissions growth slows somewhat from its current pace. Every three decades, the earth will warm more than it has in the past century. The temperature over much of the inland continental United States will likely rise nearly 1°F per decade (and in Alaska even faster). This unprecedented rate of temperature change could continue for decades.

The first brutal impacts will be marathon heat waves that last for weeks over many states. Americans have not experienced this type of extreme extended heat wave, but Europe did in August 2003. The oppressive heat brought temperatures in the upper 90s or higher across much of the continent for three weeks, and killed 15,000 people in France, 7,000 in Germany, 4,200 in Italy, and more than 2,000 in Great Britain, which on August 10 recorded its first-ever temperature over 100°F.

Scientists have studied this torrid heat wave extensively. A 2004 study in *Nature,* by British scientists from Oxford University and the Hadley Centre for Climate Prediction and Research, examined the role of greenhouse gas emissions. It concluded that human influence more than doubled the risk of such a deadly heat wave. If we stay on our current emissions trajectory, more than half of European summers will be hotter than 2003 within the next four decades. By the end of the century, "2003 would be classed as an anomalously *cold* summer relative to the new climate," the study notes.

Particularly worrisome will be shortages of water, which is essential to human life and agriculture. And large parts of the world already suffer water shortages. Moreover, many proposed solutions to our energy needs, including biofuels and hydrogen production, require huge quantities of water.

THE PRESENT IS PROLOGUE

To see what is likely to happen during Planetary Purgatory, let's look at what has happened already. Since the 1970s, the number of "very dry areas" on the planet, as defined by the widely used Palmer Drought Severity Index, has more than doubled, to about 30 percent of the global land. As a major study by the National Center for Atmospheric Research concluded, "These results provide observational evidence for the increasing risk of droughts as anthropogenic [human caused] global warming progresses and produces both increased temperatures and increased drying."

Not surprisingly, but rarely reported in context, wildfires have been on the rise worldwide for half a century. Every decade since the 1950s has seen an increase in major wildfires in the United States and around the world.

Large parts of the country have been getting hotter and drier, and suffering extended droughts. "The period since 1999 is now officially the driest in the 98 years of recorded history of the Colorado

River, according to the United States Geological Survey," noted a 2004 *New York Times* article. In March 2006, Phoenix set a record with more than 140 consecutive rainless days. "The average temperature for the continental United States from January through June 2006 was the warmest first half of any year since records began in 1895," reported NOAA's National Climatic Data Center. In June, 45 percent of the contiguous United States was in a moderate-to-extreme state of drought. By July, the figure was 51 percent.

Although the 2005 wildfire season, which ravaged 8.5 million acres, was record-breaking, the record it broke was from 2000, when wildfires consumed 8.25 million acres. From 2000 through 2005, wildfires destroyed nearly 30 million acres, some 47,000 square miles—an area *the equivalent of Pennsylvania.* Stunningly, 2006 has already broken the record set in 2005, with 8.7 million acres burned by mid-September.

Not only do drought and high temperatures increase the number of wildfires, they also lead to a greater range of pests that feast on trees whose defenses have been weakened by heat and lack of water. Trees from the Southwest up to Alaska are dying by the millions.

A 2005 study led by the University of Arizona, with the Los Alamos National Laboratory and the U.S. Geological Survey, examined a huge 3-million-acre die-off of vegetation in 2002–2003 "in response to drought and associated bark beetle infestations" in the Four Corners area (Arizona, New Mexico, Colorado, and Utah). This drought was not quite as severe as the one that region experienced in the 1950s, but it was much warmer, hence it fit the global-warming model. The recent drought had "nearly complete tree mortality across many size and age classes," whereas "most of the patchy mortality in the 1950s was associated with trees [more than] 100 years old."

Most of this tree death was caused by bark beetle infestation, and "such outbreaks are tightly tied to drought-induced water stress." Healthy trees defend themselves by drowning the tiny pine beetles in resin. Without water, weakened, parched trees are easy meals for bugs.

"We're seeing changes in [mountain pine beetle] activity from Canada to Mexico," said Forest Service researcher Jesse Logan in July 2004, "and the common thing is warming temperatures." According to the Department of Forest Resource Management at the University of British Columbia, the beetle infestation has spread to higher and more northern regions thanks in large part to climate change. And milder winters since 1994 have reduced the winter death rate of beetle larvae in Wyoming from 80 percent per year to under 10 percent.

In a February 2006 speech on climate change, Senator Lisa Murkowski of Alaska pointed out that the tremendous recent warming had opened the door to the "voracious spruce bark beetle," which devastated more than 3 million acres in Alaska, "providing dry fuel for outbreaks of enormous wild fires." Half of the wildfires in the record-breaking 2005 season were in Alaska.

I have been focusing on U.S. impacts, but the grim reality in British Columbia is too stunning to ignore. The Canadian and British Columbia Forest Service have reported that as of 2004, the mountain pine beetle infestation had killed 280 million cubic meters (10 billion cubic feet) of stately British Columbia pine trees, of which 170 million cubic meters would have been harvestable. By 2014, they project the beetle will have killed 80 percent of the harvestable pine trees—more than 800 million cubic meters. By 2025, virtually all may be gone over a region the size of North Dakota or Washington State. That is especially likely now that "it has become apparent that B.C. is facing the 'worse-case scenario,'" according to the University of British Columbia. So Canada will now log the pines as fast as possible: "Harvest levels in the region will be increased significantly over the next decade." Even so, the infestation may well spread, and then "forests across Canada may be at risk."

The authors of the 2005 study on vegetation die-off warn that the recent drought in the Four Corners "may be a harbinger of future global-change-type drought throughout much of North America and elsewhere, in which increased temperatures in concert with multidecadal drought patterns" cause unprecedented changes in ecosystems. In 2005 climatologist Jonathan Overpeck noted that this study, together with the recent evidence that temperature and annual precipitation are headed in opposite directions, raises the question of whether we are at the "dawn of the *super-interglacial drought.*"

The increased risk of severe drought we are seeing today was predicted back in 1990 by scientists at NASA's Goddard Institute of Space Studies. Their model also suggested that, in the second half of this century, severe drought, which was already occurring with about 5 percent frequency by 1990, will occur *every other year*—and more frequently in the West. The huge population growth in the western United States during the twentieth century happened to coincide with relatively wet weather in the region, weather that will likely prove to be an anomaly. One 2004 newspaper article noted, "The development of the modern urbanized West—one of the biggest growth spurts in the nation's history—may have been based on a colossal miscalculation."

Global warming also reduces the snowpack, and "snow is our water storage in the West," notes Philip Mote, climatologist for the state of Washington. States such as Montana see only 18 inches of precipitation a year. Portland gets 36 inches a year, but only one-tenth of that is during the summer. Snowmelt comprises 75 percent of all water in western streams. The warming of the last few decades has already reduced snowpacks at five out of six western snow-measurement sites. Many have suffered a 15 to 30 percent decline. And warming has moved up the peak of the annual spring runoff. In California's Sierra Nevada, streams peak as much as three weeks earlier than they did only a few decades ago.

By midcentury, warming is likely to reduce western snowpacks *by up to 60 percent* in regions such as the Cascade Range of Oregon and Washington. Summertime stream flows are projected to drop 20 to 50 percent. By century's end, the Cascades might be snow-free by April 1, and western streams might peak two months earlier than they once did. This will inevitably lead to more wildfires. A 2006 study led by the Scripps Institution of Oceanography found that the greatest increases in wildfires since 1970 were associated with warmer temperatures and earlier snowmelts, which reduce humidity and expose forests to the full effect of arid summers.

What will wildfires be like during the Planetary Purgatory era and beyond? The 2006 Scripps study compared the period 1987–2003 with the period 1970–1986. The researchers found that the active wildfire season in the West has increased 78 days and that major fires now burn 37 days—nearly five times as long as they did in the first period. And yet the average spring and summer western temperatures rose only 0.87°C (1.6°F) from the earlier period to the recent one. With current emissions trends, the West is likely to see June–August temperatures rise between 2°C and 5°C over the next half-century—suggesting we can expect a dramatic increase in fires.

Researchers at the U.S. Forest Services Pacific Wildland Fire Sciences Lab looked at past fires in the West to create a statistical model of how future climate change may affect wildfires. Their work suggests that "the area burned by wildfires in 11 Western states could double . . . if summer climate warms by slightly more than a degree and a half" centigrade. On our current emissions path, this is likely to happen by midcentury. By century's end, states such as Montana, New Mexico, Washington, Utah, and Wyoming could see burn areas increase *five times.*

If we don't change course soon, the West faces a scorching climate—Hell and No Water—with summers that are far hotter and drier, longer wildfire seasons with more ferocious fires, and, at the same time, far less water for agriculture and hydropower.

THE NEED FOR SYSTEMS THINKING

Global warming is so challenging and so potentially devastating because it is a systems problem. Although the basic definition of a system is simple—"any set of interconnected elements"—many systems, such as our climate, are exceedingly complicated.

The word *environment* comes from Old French, *viron*, meaning "circle." Since the word *cycle* also derives from *circle*, let's call the environment the cycle of life. I have not centered this book on the environment per se—on the destruction of the coral reefs or the threat to the polar bears—because so many good books have already done so and because my focus here is on the risk to the health and well-being of current and future generations of Americans.

I am a physicist who has studied and written about systems. Systems are dominated by unexpected and nonintuitive behavior because they have feedbacks, thresholds, delays, and nonlinearities. To understand the climate system, it is critical to recognize the distinction between atmospheric *concentrations* of CO_2 (the total stock of CO_2 already in the air) and annual *emissions* of CO_2 (the yearly new flow into the air). A 2002 study led by John Sterman, director of the System Dynamics Group at the MIT Sloan School of Management, found that even "highly educated graduate students" held many myths about the climate system.

> Many believe temperature responds immediately to changes in CO_2 emissions or concentrations. Still more believe that stabilizing emissions near current rates would stabilize the climate, when in fact emissions would continue to exceed removal, increasing GHG [greenhouse gas] concentrations and [planetary heating]. Such beliefs support "wait and see" policies, but violate basic laws of physics.

In fact, until annual carbon dioxide *emissions* drop to about one-fifth of current levels, *concentrations* of heat-trapping carbon dioxide will continue to rise, and with rising *concentrations,* the pace of climate change will continue to accelerate.

During the Planetary Purgatory era, the painful reality of global warming will touch the lives of all Americans. We will be forced to begin a desperate scramble, together with other nations, to stop the planet's temperature rise before the Greenland Ice Sheet melts. All Americans will become expert on both annual CO_2 emissions and total atmospheric CO_2 concentrations—two quantities that will ultimately determine the fate of the next fifty generations of Americans. I predict they will eventually be reported with as much fanfare as the gross domestic product.

As an important aside, scientists and government agencies often use carbon, C, rather than carbon dioxide, CO_2, as a metric. Carbon dioxide is the greenhouse gas. Carbon is found in fossil fuels and soils and trees. The global carbon cycle is what many scientists study. You need familiarity with both quantities to follow the scientific and political debates about climate science and climate solutions. The key relationship to remember is:

1 ton carbon, C, equals 3.67 tons carbon dioxide, CO_2

Thus 11 tons of carbon dioxide equals 3 tons of carbon, and a price of $30 per ton of carbon dioxide equals a price of $110 per ton of carbon.

In 2005, fossil fuel combustion released into the air more than 26 billion tons of CO_2 (more than 7 billion tons of carbon). This is five times the annual rate of emissions from the 1940s. For the past decade, annual emissions have been rising about 2 percent per year, in large part driven by China and the United States. This rate of growth seems likely to continue through 2015 and possibly through

2025, barring a sudden reversal of U.S. (and Chinese) climate and energy policy. In 2005, the U.S. Department of Energy forecast that global annual emissions would exceed 30 billion tons of CO_2 in 2010 and, in 2025, 38 billion tons of CO_2 (more than 10 billion tons of carbon). Such rapid *emissions* growth by 2025 would make *concentrations* soar and take the nation and the world to the very edge of catastrophe.

While emissions might be thought of as the water flowing into a bathtub, atmospheric concentrations are the water level in the bathtub. Emissions are analogous to the federal budget deficit we incur each year, and concentrations are analogous to the total national debt that has been accumulated.

In 2005, atmospheric concentrations of carbon dioxide were 380 parts per million, about a third higher than the preindustrial average of about 280 ppm. In recent years, the rate of growth of concentrations has doubled. Concentrations are now climbing more than 2.5 ppm a year. By 2025, concentrations are projected to be 420–430 ppm. During Planetary Purgatory, concentrations are projected to rise an average of 3 ppm a year. In this scenario, by 2050, atmospheric concentrations would hit 500 ppm. Yet once we get much past 500 ppm, the complete melting of the Greenland ice sheet and the resulting 20-foot sea-level rise become all but inevitable.

Now it begins to be clear how desperate we will be in Planetary Purgatory. Suppose that America takes no serious action on climate while George W. Bush is president, and we successfully block any serious efforts by other nations. Suppose that then, starting about 2010, we take some wishy-washy actions to slow our emissions growth, while China and other developing nations continue their booming growth largely unchecked (thanks to growing populations, industrialization, and a rapidly expanding middle class). Suppose we continue making modest investments in developing new technology. Near 2020, America starts to get more serious, and we organize international commitments that slow global emissions growth *by half.*

Finally, in 2025, the entire world wakes up to the full gravity of global warming. Now we adopt the aggressive five-decade effort to deploy the best existing energy technology described in chapter 1 (modified from the analysis by Princeton's Stephen Pacala and Robert Socolow). From 2025 through 2075, the world achieves eight remarkable changes:

1. We launch a massive performance-based efficiency program for homes, commercial buildings, and new construction.
2. We launch a massive effort to boost the efficiency of heavy industry and expand the use of cogeneration (combined heat and power).
3. We capture the CO_2 from 800 new large coal plants and store it underground.
4. We build 1 million large wind turbines (or the equivalent in renewables like solar power).
5. We build 700 new large nuclear-power plants while shutting down no old ones.
6. We require every car to have an average fuel economy of 60 mpg.
7. We enable every car to run on electricity for short distances (requiring another half-million large wind turbines) before reverting to biofuels (requiring one-twelfth the world's cropland).
8. We stop all tropical deforestation, while doubling the rate of new tree planting.

Pacala and Socolow call these "wedges," since each starts slowly but then rises in impact over the 50 years and ultimately avoids the emission of 1 billion tons of carbon per year.

Had we started these eight wedges in 2010, global carbon emissions would have remained frozen at 8 billion metric tons per year. But because we delayed, because we started in 2025, they will merely *slow* emissions growth, so that global carbon emission will rise from 10 billion metric tons per year in 2025 to 12 billion metric tons per year in 2075. Finally, suppose that, starting in 2075, we adopt even more aggressive use of advanced energy technologies, and global emissions actually start *dropping* 1.5 percent per year.

In this scenario, carbon dioxide concentrations would exceed 600 ppm in 2100—and perhaps exceed 750 ppm, given the likely effect of the climate system's vicious cycles, as we will see shortly—and continue to rise. The temperature rise from current levels to 2100 would be a whopping 2.5°C or more. The outcome: We caused an eventual 20-foot sea-level rise, and we probably caused an eventual 80-foot rise. We didn't prevent a century or more of super-hurricanes and mega-droughts. We were insufficiently desperate and poorly led. We waited for new technology to show up in 2025 instead of deploying existing technology at once.

And if we do wait until 2025, the relatively painless technology-driven solutions that are available in 2007 or 2010 will no longer be sufficient to avoid climate catastrophe. Our actions will have to be far more desperate and aggressive. Just how desperate and aggressive critically depends on the myriad feedback loops in the climate system that will almost certainly punish any unwise delay in taking global warming seriously.

CLIMATE REALITY VERSUS CLIMATE MODELS

The earth's climate system is "far from being self-stabilizing," in the words of climatologist Wallace Broecker, but is "an ornery beast which overreacts even to small nudges." Push it too hard in one direction, you get an ice age, in another direction, you get 80-foot-higher sea levels. This suggests that the climate system has one or more vicious cycles, in which a little warming causes a change that speeds up warming, as when warming melts highly reflective Arctic ice, replacing it with the blue sea, which absorbs far more sunlight and hence far more solar energy, causing the Arctic Ocean to heat up more, melting more ice, and so on. Vicious cycles are often called "positive feedbacks" in the scientific literature, because these feedbacks add to and increase the effect. It is not a term I will use much here because it has a positive connotation in general usage— everybody wants to get positive feedback—whereas everybody should want to avoid the vicious cycles of the climate system.

The models that tell us how much warming we will get from a certain level of carbon dioxide emissions do not fully account for all of the vicious cycles. Thus, these models almost certainly significantly *underestimate* the climate's likely response to our emissions of greenhouse gases, a view shared by a number of recent studies and most of the climate scientists I talked to, such as Harvard's Dan Schrag. Let's look briefly at three studies from 2006.

Scientists analyzed data from a major expedition to retrieve deep marine sediments beneath the Arctic to understand the Paleocene-Eocene Thermal Maximum, a brief period some 55 million years ago of "widespread, extreme climatic warming that was associated with massive atmospheric greenhouse gas input." This study, published in *Nature,* found Artic temperatures almost beyond imagination—above 23°C (74°F)—temperatures far warmer than current climate models had predicted when applied to this period. The three dozen authors conclude that existing climate models are missing crucial factors that can significantly amplify polar warming.

A second study looked at temperature and atmospheric changes during the Middle Ages. The study found that the effect of vicious cycles in the climate system—where global warming boosts atmospheric CO_2 levels—"will promote warming by an extra 15 percent to 78 percent" compared with typical estimates by the U.N.'s Intergovernmental Panel on Climate Change. The study notes that these results may even be conservative because they ignore other greenhouse gases such as methane, whose levels will likely be boosted as temperatures warm.

The third study looked at temperature and atmospheric changes during the past 400,000 years. It found evidence for significant increases in both CO_2 and methane (CH_4) levels as temperatures rise. The conclusion: If our current climate models correctly accounted for such vicious cycles, "we would be predicting a significantly greater increase in global warming than is currently forecast over the next century and beyond"—as much as 1.5°C warmer this century alone.

Let's look at some key vicious cycles that climate modelers are missing or underestimating.

THE FOUR (POTENTIAL) SOURCES OF THE APOCALYPSE

For the last few decades, nearly 60 percent of the carbon dioxide that we have been adding to the atmosphere has stayed there. Where did the rest go? The other 40 percent has been absorbed by several "sinks"—the ocean, soils (including permafrost), and vegetation. They are called sinks because they absorb carbon and remove it from the ecosystem. Returning to the bathtub analogy, a carbon sink is just like the drain in your bathtub. The sources, including cars, factories, and power plants, are like faucets. As long as the sources generate more carbon dioxide than the sinks can drain, atmospheric concentrations (the water level in the bathtub) will continue to rise.

This is called the *global carbon cycle*. At some threshold of carbon dioxide concentrations and temperature rise, most scientists believe that one or all of these sinks will saturate—like clogged-up drains, they will not be able to absorb any more. Some carbon sinks may actually turn into *sources* of greenhouse gases. Preliminary evidence suggests that may be starting to happen already. I think we will know for certain by the 2025–2050 era. Let's look at four key sinks that could drive vicious cycles: oceans, soils, permafrost, and vegetation.

First, the oceans. According to a 2005 report by the United Kingdom's Royal Society, ocean warming leads to a "decreased mixing between the different levels in the oceans." That, in turn, "would reduce CO_2 uptake, in effect, reducing the oceanic volume available to CO_2 absorption from the atmosphere." In other words, if surface water that has absorbed CO_2 does not switch places with deeper water, the ocean will absorb less and less CO_2 over time and more will stay in the atmosphere. The increased ocean stratification would also tend to separate some phytoplankton from their nutrients, "leading to a decline in oceanic primary production," which would also reduce the ocean's ability to take up carbon, which means more CO_2 would stay in the air, and on and on. Finally, on our current CO_2 emissions trend, the ocean will become so acidic that coral reefs and other sea life will be devastated, further reducing the ocean's ability to absorb carbon.

Second, the soils. Warming can cause soils to stop taking up CO_2 and, ultimately, to start releasing it. A 2002 study of Texas grasslands found that as CO_2 concentrations increase, the ability of the soil to take up carbon slowed more rapidly than expected, "indicating that we are currently at an important threshold." The study notes that "the ability of soils to continue as sinks is limited." British soil experts have been monitoring their soil at several thousand sites in England and Wales since 1978. In 2005, they reported that the soils are releasing their carbon. The net carbon content has been dropping 0.6 percent per year—a huge amount considering that the

CO_2 released from British soils would be enough to erase the industrial-emissions reductions the country has achieved so far with its enlightened energy policies.

Third, the tundra, Arctic permafrost, and frozen peat. The permafrost is soil that stays below freezing (0°C or 32°F) for at least two years. Peat is basically mulch, or organic matter that is partially decomposed. It is found around the globe, but it is frozen near the poles. Normally, plants capture carbon dioxide from the atmosphere during photosynthesis and slowly release that carbon back into the atmosphere after they die. But the Arctic acts like a freezer, and the decomposition rate is very low. So frozen peat is "a locker of carbon," as UCLA scientist Laurence Smith explained at an American Meteorological Society seminar in February 2006.

How much? According to a June 2006 *Science* article by Russian and American scientists, nearly 1,000 billion metric tons of carbon (some 3,600 billion metric tons of carbon dioxide) are locked up in the Arctic's permafrost. That exceeds all the carbon dioxide currently in the atmosphere. The permafrost may contain more than a third of all carbon stored in soils globally, much of it in the form of methane. The problem: Global warming is melting the top layer of permafrost, creating the possibility of large releases of soil carbon, and that is a potentially devastating vicious cycle. We are defrosting the tundra freezer—and at an unprecedented rate.

A 2006 study by Alaska researchers finds rapid degradation to key elements of the permafrost "that previously had been stable for 1000s of years." The study, titled "Abrupt Increase in Permafrost Degradation in Arctic Alaska," concludes that this recent degradation exceeds changes seen earlier in the twentieth century by a factor of ten to a hundred.

New Scientist magazine reported in August 2005 that in western Siberia a frozen peat bog the size of France and Germany combined is turning into "a mass of shallow lakes," some almost a mile wide. In the past 40 years, the region has warmed by 3°C, greater warming than almost anywhere else in the world, in part because of the vicious cycle described earlier: Warming melts highly reflective ice and replaces it with dark soils, which absorb more sunlight and warm up, melting more ice, and on and on.

Russian botanist Sergei Kirpotin describes an "ecological landslide that is probably irreversible and is undoubtedly connected to climatic warming." The entire western Siberian sub-Arctic region is melting, and it "has all happened in the last three or four years," according to Kirpotin, who believes we are crossing a critical threshold. The peat bogs formed near the end of the last ice age some 11,000 years ago. They generate methane, which, up until now, has mostly been trapped within the permafrost, and in even deeper ice-like structures called clathrates. The Siberian frozen bog is estimated to contain 70 billion tons of methane (CH_4). If the bogs become drier as they warm, the methane will oxidize and the emissions will be primarily CO_2. But if the bogs stay wet, as they have been recently, the methane will escape directly into the atmosphere.

Either way we have a dangerous vicious cycle, but the wet bogs are worse because methane has twenty times the heat-trapping power of carbon dioxide. Some 600 *million* metric tons of methane are emitted each year from natural and human sources, so if even a small fraction of the 70 *billion* tons of methane in the Siberian bogs escapes, it will swamp those emissions and dramatically accelerate global warming. Researchers monitoring a single Swedish bog, or mire, found it had experienced a 20 to 60 percent increase in methane emissions between 1970 and 2000. In some methane hot spots in eastern Siberia, "the gas was bubbling from thawing permafrost so fast it was preventing the surface from freezing, even in the midst of winter."

Even if the tundra carbon is all emitted as carbon dioxide instead of methane, the consequences would be disastrous. Carbon emissions from human activity already exceed 7 billion tons a year, and we are on track to be at 10 billion tons a year by 2025. But as we have already seen, if we exceed annual emissions levels of 10 billion tons for any significant length of time, we will have no chance of avoiding catastrophic warming.

A major 2005 study led by NCAR climate researcher David Lawrence found that virtually the entire top 11 feet of permafrost around the globe could disappear by the end of this century. Using the first "fully interactive climate system model" applied to study permafrost, the researchers found that if we somehow stabilize carbon dioxide concentrations in the air at 550 ppm, permafrost would plummet from more than 4 million square miles today to 1.5 million. If concentrations hit 690 ppm, permafrost would shrink to just 800,000 square miles.

While these projections were done with one of the most sophisticated climate-system models in the world, the calculations *do not yet include the feedback effect of the released carbon from the permafrost.* That is to say, the CO_2 concentrations in the model rise only as a result of direct emissions from humans, with no extra emissions counted from soils or tundra. Thus they are conservative numbers—or *overestimates*—of how much CO_2 concentrations have to rise to trigger *irreversible* melting.

David Lawrence told me that NCAR's climate model will not incorporate these feedbacks for many years. And most major climate models do not include these crucial feedbacks (one exception is below). Thus, the *Fourth Assessment Report* by the Intergovernmental Panel on Climate Change, coming out this year (2007), almost certainly *underestimates* greenhouse gas forcings and climate change this century. In short, we have a much tougher task than the U.N.'s consensus-based process has been telling us.

By the end of Planetary Purgatory, most of the tundra may be unsavable.

Fourth, the tropical forests. Tropical forests store carbon, and destroying them releases that carbon. Intact tropical forests serve as a carbon sink for slightly more than 1 billion metric tons of carbon a year. A 2006 article by British scientists reviewing the current state of knowledge on tropical forests and carbon dioxide estimated that tropical deforestation released emissions "at the higher end" of the reported range of 1 to 3 billion metric tons of carbon a year.

Unfortunately, we do not appear prepared to stop current deforestation trends, while the carbon sink is likely to shrink because of increased drought, wildfires, and temperatures. The mechanisms are deadly enough individually, but when they interact synergistically the effects are multiplied and create a classic vicious cycle.

We've already seen how high temperatures and drought have combined to create record wildfires in the United States, but the situation is far worse in other parts of the world. The global fires of 1997–1998 "may have released carbon equivalent to 41 percent of worldwide fossil fuel use," according to a 2003 *Nature* article. Over Southeast Asia and Latin America alone, acreage equal to half of California burned out of control. While Indonesia lost more than double the acreage the United States lost in its record-breaking 2005 wildfire season, that developing country spends only about 2 percent of what we do on fire suppression. The article concluded grimly, "Pan-tropical forest fires will increase as more damaged, less fire-resistant, forests cover the landscape."

In Indonesia, both rain forests and peat lands burned. Carbon-rich tropical peat deposits can be more than 60 feet deep. A 2002 *Nature* article reported, "The extensive fire damage caused in 1997 has accelerated changes already being caused in tropical peatlands by forest clearance and drainage." Using satellite images to compare logging activity for the years 1997 and 2000, the authors found that "logging had increased by 44 percent," which made the remaining forests "more susceptible to fire in the future." Absent a major effort to address the problem, "tropical peatlands will make a significant contribution to global carbon emissions for some time to come."

In 2005, the Amazon was suffering a brutal drought—in many regions the harshest since records began a hundred years ago. By October, the governor of Amazonas State had declared a "state of public calamity." The threat to the Amazon forest is grave. The Woods Hole Research Center in Santarém on the Amazon River reported in 2006 that the "forest cannot withstand more than two consecutive years of drought without breaking down." Dr. Dan Nepstad of Woods Hole expects "mega-fires" to sweep across the jungle if it gets too dry.

Today, about 20 percent of the rain forest has been chopped down, and another 22 percent has been hurt enough by logging that sufficient sunlight can reach the forest floor to dry it out. Models suggest that when 50 percent of the forest is destroyed—which some models project for 2050—it will have crossed a "tipping point" beyond which its destruction cannot be stopped. In the coming decades, drought and heat will combine to devastate the rain forest and its canopy, reducing local rainfall and further accelerating the drought and local temperature rise, ultimately causing the release into the atmosphere of huge amounts of carbon currently locked in Amazon soils and vegetation, another fearsome feedback loop at work.

CROSSING THE POINT OF NO RETURN

Global warming is on the verge of dramatically transforming the global carbon cycle, causing the release of carbon from some soils, tundra, and forests, while slowing the uptake of carbon by the ocean and other carbon sinks.

The United Kingdom's Hadley Centre for Climate Prediction and Research has one of the few climate models that incorporates a significant number of carbon-cycle feedbacks, particularly in soils and tropical forests. In a 2003 study, they found that a typical fossil fuel emissions scenario for this century, which would have led to carbon dioxide concentrations in 2100 of about 700 ppm *without* feedbacks, led instead to concentrations of 980 ppm *with* feedbacks, a huge increase. Even ignoring feedbacks, keeping concentrations below 700 ppm requires the United States and the world to start slowing carbon dioxide emissions from coal, oil, and natural gas significantly by 2015 and to stop the growth almost entirely after 2025.

In 2006 the Hadley Centre, working with other British researchers, published an important study, "Impact of Climate-Carbon Cycle Feedbacks on Emissions Scenarios to Achieve Stabilisation," which included both ocean and terrestrial carbon-cycle feedbacks (though they do not specifically model carbon emissions from defrosting tundra). The study found that such feedbacks reduce the amount of fossil fuel emissions we can release by 21 percent to 33 percent.

We have no room for error. The Hadley study finds that just to stabilize at 650 ppm, annual emissions this century will have to *average* under 9 billion tons of carbon, a level that emissions will probably achieve by 2015. Absent the feedbacks, annual emissions this century could have averaged nearly a third more.

There appears to be a threshold beyond which it becomes more and more difficult for us to fight the feedbacks of the carbon cycle with strong energy policies that reduce fossil fuel emissions into the air. While the threshold is not known precisely today, it appears to be somewhere between 450 ppm and 650 ppm, based on my review of the literature and conversations with climate scientists. By 2025, we'll know much better where it is. Unfortunately, on our current path, the world's emissions and concentrations will be so high by 2025 that the "easy" technology-based strategy will not be able to stop us from crossing the very high end of the threshold range.

That's why I am calling the second quarter of this century Planetary Purgatory. Barring a major reversal in U.S. policies in the very next decade, come the 2020s, most everyone will know the grim fate that awaits the next fifty generations. But the only plausible way to avoid it will be a desperate effort to cut global emissions by 75 percent in less than three decades—a massive, sustained government intervention into every aspect of our lives on a scale that far surpasses what this country did during World War II. That would indeed be punishment for our sins of inaction.

Failing that desperate effort, we would end up at midcentury with carbon emissions far above current levels, and concentrations at 500 ppm, rising 3 to 4 ppm a year—or even faster if the vicious cycles of the climate system have kicked in.

We have passed the point of no return.

CHAPTER FOUR

2050–2100: HELL AND HIGH WATER

We could get a meter [of sea-level rise] easy in 50 years.
    —Bob Corell, chair, Arctic Climate Impact
        Assessment, 2006

The peak rate of deglaciation following the last Ice Age was . . . about one meter [39 inches] of sea-level rise every 20 years, which was maintained for several centuries.
    —James Hansen, director, Goddard Institute for
        Space Studies (NASA), 2004

Sea-level rise of 20 to 80 feet will be all but unstoppable by mid-century if current emissions trends continue. The first few feet of sea-level rise alone will displace more than 100 million people worldwide and turn all of our major Gulf and Atlantic coast cities into pre-Katrina New Orleans—below sea level and facing super-hurricanes.

How fast can seas rise? For the past decade, sea levels have been rising about 1 inch a decade, double the rate of a few decades ago. The *Third Assessment Report* of the U.N. Intergovernmental Panel on Climate Change (IPCC), released back in 2001, projected that sea levels would rise 12 to 36 inches by 2100, with little of that rise coming from either Greenland or Antarctica. Seas rise mainly because ocean water expands as it gets warmer, and inland glaciers melt, releasing their water to the oceans.

Sea-level rise is a lagging indicator of climate change, in part because global warming also increases atmospheric moisture, as we've seen. More atmospheric moisture probably means more snowfall over both the Greenland and Antarctica ice sheets, which would cause them to gain mass in their centers even as they lose mass at the edges. Until recently, most scientists thought that the primary mechanism by which these enormous ice sheets would lose mass was through simple melting. The planet warms and ice melts—a straightforward physics calculation and a very slow process, with Greenland taking perhaps a thousand years or more to melt this way, according to some models.

Since 2001, however, a great many studies using direct observation and satellite monitoring have revealed that both of the two great ice sheets are losing mass at the edges much faster than the models had predicted. We now know a number of physical processes can cause the major ice sheets to disintegrate faster than by simple melting alone. The whole idea of "glacial change" as a metaphor for change too slow to see will vanish in a world where glaciers are shrinking so fast that you can actually watch them retreat.

The disintegration of the Greenland and Antarctic ice sheets is a multistage process that starts with the accelerated warming of the Arctic.

THE END OF THE ARCTIC AS WE KNOW IT

Global warming tends to occur faster at high latitudes, especially in the Arctic. That is called polar amplification. Arctic warming is amplified for several synergistic reasons, as explained in the most comprehensive scientific survey completed to date, the December 2004

Arctic Climate Impact Assessment, by leading scientists from the eight Arctic nations—Canada, Denmark/Greenland, Finland, Iceland, Norway, Russia, Sweden, and the United States:

1. Warming melts highly reflective white ice and snow, which is replaced by the dark blue sea or dark land, both of which absorb far more sunlight and hence far more solar energy.
2. In the Arctic, compared with lower latitudes, "more of the extra trapped energy goes into warming rather than evaporation."
3. In the Arctic, "the atmospheric layer that has to warm in order to warm the surface is shallower."
4. So, when the sea ice retreats, the "solar heat absorbed by the oceans in summer is more easily transferred to the atmosphere in winter."

And this leads to more snow and ice melting, further decreasing Earth's reflectivity (albedo), causing more heating, which the thinner Arctic atmosphere spreads more quickly over the entire polar region, and so on and so on.

We can witness this classic feedback loop today at the North Pole, where the summer ice cap has shrunk more than 25 percent from 1978 to 2005, a loss of 500,000 square miles of ice, an area twice the size of Texas. The Arctic winters were so warm in both 2005 and 2006 that sea ice did not refreeze enough to make up for the unprecedented amount of melting during recent summers. A synthesis report in August 2005 by twenty-one leading climate scientists, supported by the U.S. National Science Foundation's Arctic Systems Science Program, described the future in terms that were unusually stark for a group of scientists:

At the present rate of change, a summer ice-free Arctic Ocean within a century is a real possibility, *a state not witnessed for at least a million years.* The change appears to be driven largely

by *feedback-enhanced global climate warming,* and there seem
to be few, if any, processes or feedbacks within the Arctic system that are capable of altering the trajectory toward this
"super interglacial" state. [Emphasis added.]

We appear to be crossing a threshold in the Arctic, one that existing models did not predict would happen so fast. "The recent sea-
ice retreat is larger than in any of the (19) IPCC [climate] models,"
Tore Furevik pointed out in a November 2005 talk on climate-system
feedbacks. He is deputy director of Norway's Bjerknes Centre for
Climate Research. Once again, the models on which the IPCC bases
its conclusions appear to be "too conservative," either underestimating or missing entirely relevant climate feedbacks. Most models suggest that the Arctic Ocean will see ice-free summers by 2080 to 2100.
At our current pace, this will happen long before then.

According to a 2005 *Science* article, key Arctic landmasses have
warmed "0.3° to 0.4°C per decade since the 1990s," double the rate
of the previous two decades. A 2005 study led by the Institute of Arctic Biology at the University of Alaska at Fairbanks and the U.S.
Geological Survey, estimated that the reduced snow cover and albedo in the summertime Arctic landscape, caused by global warming, added local atmospheric heating comparable to what a doubling
of CO_2 levels (to 550 ppm) would do over many decades to the global
atmosphere. In short, the dramatic climatic changes in the Arctic
today are a warning to us of both the pace and degree of change
America will experience early in the second half of this century.

The study also noted that "the continuation of current trends
in shrub and tree expansion could further amplify this atmospheric
heating by two to seven times." As the permafrost thaws, creating a
moist, nutrient-rich environment for vegetation, polar amplification will accelerate. We have very few climate models that incorporate the impact of such changes in vegetation, which again indicates
how likely it is that we are underestimating the future warming of

the Arctic. The scientific evidence is simply accumulating too fast to model adequately.

New research suggests that the summer Arctic could be ice-free far sooner than anyone ever imagined. Simply looking at the shrinking *area* of the Arctic ice misses an even more alarming decline in its thickness and hence its *volume*. At a May 2006 seminar sponsored by the American Meteorological Society, Dr. Wieslaw Maslowski of the Oceanography Department at the Naval Postgraduate School reported that models suggest that the Arctic lost one-third of its ice volume from 1997 to 2002. He made an alarming forecast: "If this trend persists *for another 10 years*—and it has through 2005—we could be ice-free in the summer" (emphasis added).

The loss of Arctic ice has little effect on sea levels because the ice is floating on the Arctic Ocean. Like a floating ice cube in a glass of water, when it melts, it doesn't change the water level. Why, then, should we be worried? Because in the Arctic, the accelerating warming of the land, air, and ocean sets the stage for one of the severest impacts of climate change facing our country—extreme sea-level rise from the disintegration of the Greenland Ice Sheet.

THE END OF GREENLAND—AND COASTAL LIFE— AS WE KNOW IT

> *Models indicate that warming over Greenland is likely to be of a magnitude that would eventually lead to a virtually complete melting of the Greenland ice sheet, with a resulting sea-level rise of about seven meters (23 feet).*
> —Arctic Climate Impact Assessment, 2004

The Greenland Ice Sheet extends over some 1.7 million square kilometers (more than 650,000 square miles). It is as large as the entire state of Alaska and almost as big as Mexico. It is 3 kilometers (nearly 2 miles) at its thickest. It contains nearly 3 million cubic kilometers (750,000 cubic miles) of ice. Unlike the Arctic ice cap, Greenland's landlocked ice, when it returns to the ocean, causes sea levels to rise.

Current climate models project that the entire ice sheet will melt if Greenland warms only about 4.5°C (8.1°F). Since Greenland is currently warming much faster than the planet as a whole, that is likely to occur when the planet warms more than 3°C compared with levels of the late 1800s. Exceeding such warming by 2100 is a near certainty if greenhouse gas concentrations significantly exceed 550 parts per million, a doubling from preindustrial levels. On our current path, we may hit 550 by midcentury.

Once the warming passes this threshold, the melting may become almost unstoppable. As climatologist Jonathan Gregory has pointed out, melting lowers the altitude of the ice-cap surface, which leads to more warming and reduced snowfall, another vicious cycle. Until recently, the conventional wisdom was that Greenland would take a thousand years or more to lose its ice sheet. But that assumed that the loss in mass would come exclusively from simple melting. We now know, however, that melting is anything but simple.

A team led by NASA and MIT scientists reported in 2002 that the ice was flowing in parts of Greenland much faster during the summer melting season than the winter. They concluded that some of the water flows to the ice-bedrock interface at the bottom of the glacier and acts as a lubricant for the entire glacier to slide and glide on. This "provides a mechanism for rapid response of the ice sheets to climate change," a factor that has been given "little or no consideration in estimates of ice-sheet response to climate change."

Scientists have observed another crucial change to Greenland's glaciers in recent years—the outlet glaciers have been speeding up, thinning, and disintegrating. The Greenland Ice Sheet drains into the sea through dozens of large glaciers, although roughly half the discharge "is through 12 fast-flowing outlet glaciers, most no more than 10 to 20 kilometers across at their seaward margin, and each fed from a large interior basin of about 50,000 to 100,000 square kilometers," reported a 2006 review article in *Science*. The outlet glaciers have ice shelves or floating tongues of ice that can extend tens of kilometers past the point where the glaciers are supported by the ground. The front face of the ice shelves are hundreds of meters thick and calve or break off icebergs into the ocean.

For many years, scientists have been studying Jakobshavn Isbrae, Greenland's largest outlet glacier, which drains some 6.5 percent of the entire ice sheet's area. From 1950 to 1996, the glacier's terminal point, or calving front, was stable, fluctuating about 2.5 kilometers back and forth around its seasonal average. This multi-decadal stability may have been due to "resistance from the fjord walls and pinning points" that helped secure the outlet glacier.

The outlet glacier is like a cork in a champagne bottle—and humanity, with our ever-increasing emissions of heat-trapping gases, has been frantically shaking this bottle. So it should not be a total surprise that a study of the Jakobshavn Isbrae glacier using satellite images found that "in October 2000, this pattern [of stability] changed when a progressive retreat began that resulted in *nearly complete disintegration of the ice shelf* by May 2003." The cork popped. Freed from this barrier that had been holding it back, the glacier's speed increased dramatically to 12.6 kilometers (7.8 miles) a year. Ice discharge nearly doubled. The authors concluded that "fast-flowing glaciers can significantly alter ice-sheet discharge at sub-decadal timescales and that their response to climate change has at least the potential to be rapid."

Jakobshavn Isbrae's sudden behavior change is no random event. A 2006 study found a similar change in two East Greenland outlet glaciers—Kangerdlugssuaq and Helheim, which are about 200 miles apart. In both glaciers, "acceleration and retreat has been sudden, despite the progressive nature of warming and thinning over some years." The top surface height of Helheim dropped more than 150 feet in two years. The surface of Kangerdlugssuaq dropped more than 250 feet. The two glaciers together drain about 8 percent of Greenland's ice sheet. They have nearly doubled the ice transport to the sea from this area of Greenland, to 100 cubic kilometers a year, up from about 50. The authors conclude that "the most plausible sequence of events is that the thinning eventually reached a *threshold*, ungrounded the glacier tongues and subsequently allowed acceleration, retreat and further thinning." This represents a step change in ice dynamics "not included in current models." The authors warn that given such behavior in three disparate outlet glaciers, "we should expect further Greenland outlet glaciers to follow suit."

How fast do Greenland glaciers move these days? Using Global Positioning System equipment, researchers have clocked Helheim at speeds exceeding 14 kilometers per year, nearly triple its 2001 speed. That flow rate equals an inch a minute. In 2005, Jakobshavn Isbrae was clocked at a similar speed. You can watch these glaciers move. That isn't "glacial change"—Greenland's glaciers are moving far faster than America's climate policy.

While 2002 had been the record for surface-area melting in Greenland since 1979 (the year systematic satellite monitoring began), 2005 topped that easily. A major 2006 study led by NASA's Jet Propulsion Laboratory found that "accelerated ice discharge in the west and particularly in the east doubled the ice sheet mass deficit in the last decade from 90 to 220 cubic kilometers per year." (Los Angeles uses about 4 cubic kilometers of fresh water a year.) The study's lead author, Eric Rignot, said in 2006, "In the next 10 years, it wouldn't surprise me if the rate doubled again."

Whereas glacier acceleration was widely found below 66° north latitude between 1996 and 2000, that line had shifted to 70° north by 2005. The authors conclude, "As more glaciers accelerate farther north, the contribution of Greenland to sea-level rise will continue to increase." In short, global warming is rapidly speeding up the disintegration of the entire Greenland Ice Sheet, and if we stay on our current emissions path until the 2050–2100 era, the loss of the Greenland Ice Sheet could become irreversible, according to NASA's Jay Zwally.

The IPCC's *Third Assessment Report* in 2001, which is used as the standard by most nations for impact assessment, projected a half-meter (20-inch) sea-level rise by 2100, with a worst case of 1 meter. But that assessment assumed Greenland would contribute little to sea-level rise by 2100. The startling changes now observed in Greenland alone would suggest 20 inches is a best-case scenario for 2100—and we should plan on much worse. If glacier acceleration continues, then by itself Greenland could easily generate sea-level rise of 5 inches or more *per decade* during Hell and High Water—and for centuries to come.

Greenland is not, however, the only major ice sheet showing signs of unexpected disintegration. So is Antarctica.

THE END OF ANTARCTICA—AND CIVILIZATION— AS WE KNOW IT

> *The last IPCC report characterized Antarctica as a slumbering giant in terms of climate change. I would say it is now an awakened giant.*
>
> —Chris Rapley, head of the British Antarctic Survey, 2006

Antarctica is bigger than the United States, and its ice sheet has locked away more than *eight times* as much ice as Greenland's. It holds 90 percent of Earth's ice. As recently as the *Third Assessment Report* in 2001, many scientists were not very worried about an Antarctic contribution to sea-level rise in this century. Antarctica is 99 percent covered by ice that is on average about 2 kilometers (1.2 miles) thick. It is one huge freezer. Until recently, scientists believed that warming-induced increases in snowfall over central Antarctica would just about counterbalance whatever melting occurred along the edges.

But as with Greenland, "in the last decade, our picture of a slowly changing Antarctic ice sheet has radically altered," explained a 2005 report by the Ice Sheet Mass Balance and Sea Level committee, a group of leading climate scientists and glaciologists. As with Greenland, global warming is causing outlet glaciers to thin and disintegrate while ice flow accelerates.

The Antarctic Peninsula, which juts out in the direction of South America, is warming the fastest—about 2.0°C in the past half-century, a rate unprecedented for at least two millennia. In 2002, much of the peninsula's Larsen B Ice Shelf disintegrated in spectacular fashion. The shelf, which had probably been in existence since the end of the last ice age, lost an area larger than the state of Rhode Island in a matter of weeks. After the collapse, glaciers flowing into it sped up two- to eightfold. One glacier's surface dropped 38 meters (125 feet) in six months, leading to an additional mass loss of 27 cubic kilometers per year, just from this small part of Antarctica.

In 2005, the British Antarctic Survey and U.S. Geological Survey reported the results of a comprehensive analysis of the glaciers that drain the peninsula's ice sheet. Of 244 glaciers, 212 have retreated since the earliest positions recorded five decades ago, and they have retreated far greater distances than the few advancing glaciers have expanded. Moreover, the line of retreating glaciers has moved steadily southward during this time, toward the South Pole, suggesting the influence of global warming. The authors conclude that "the cumulative loss of ice at the fronts of these glaciers may be leading to an increased drainage of the Antarctic Peninsula that is more widespread than previously thought."

The peninsula is not the only area of Antarctica with glaciers that are warming up—and speeding up. A 2004 study in *Geophysical Research Letters* noted that over the previous decade, the grounded Amundsen Sea portion of the West Antarctic Ice Sheet has been losing 50 cubic kilometers of volume each year "due to an imbalance between snow accumulation and ice discharge." Satellite measurements reveal that the ice shelves in one major outlet glacier, Pine Island Bay, have been thinning by up to 5.5 meters per year during this time. The reason appears to be ocean currents averaging 0.5°C warmer than freezing. The Pine Island and Thwaites glaciers enter the Amundsen Sea at Pine Island Bay. They are discharging ice three times faster than they were just ten years earlier. The study concludes that "the drawdown of grounded ice shows that Antarctica is more sensitive to changing climates than was previously considered."

A major 2004 study led by NASA researchers using satellite and aircraft laser altimeter surveys found that glaciers in this sector of the ice sheet are "discharging about 250 cubic kilometers of ice per year to the ocean," much more ice than is accumulating in the areas that feed these glaciers. The glaciers are thinning far faster than they were even a decade ago. As noted, that mass loss is partly counterbalanced by increased snowfall over the rest of Antarctica, and in 2001 the IPCC projected that Antarctica would *gain* mass this century. Only three years later, the data showed otherwise.

In a surprising finding, University of Colorado at Boulder researchers reported in 2006 that Antarctica as a whole was *losing* up to 150 cubic kilometers of ice annually. They used twin satellites to measure the mass of the entire Antarctic ice sheet as part of the Gravity Recovery and Climate Experiment (GRACE). That Antarctica is rapidly losing ice was confirmed by another 2006 study, the most comprehensive survey ever undertaken of the ice sheets, led by NASA's Zwally.

Perhaps the most important, and worrisome, fact about the

West Antarctic Ice Sheet (WAIS) is that *it is fundamentally far less stable than the Greenland Ice Sheet* because most of it is grounded far below sea level. The WAIS rests on bedrock as deep as 2 kilometers underwater. The 2004 NASA-led study found that most of the glaciers they were studying "flow into floating ice shelves over bedrock up to hundreds of meters deeper than previous estimates, providing exit routes for ice from further inland if ice-sheet collapse is under way." A 2002 study in *Science* examined the underwater grounding lines—the points where the ice starts floating. Using satellites, the researchers determined that "bottom melt rates experienced by large outlet glaciers near their grounding lines are far higher than generally assumed." And that melt rate is positively correlated with ocean temperature.

The warmer it gets, the more unstable WAIS outlet glaciers will become. Since so much of the ice sheet is grounded underwater, rising sea levels may have the effect of lifting the sheet, allowing more—and increasingly warmer—water underneath it, leading to further bottom melting, more ice-shelf disintegration, accelerated glacial flow, further sea-level rise, and so on and on, another vicious cycle. The combination of global warming and accelerating sea-level rise from Greenland could be the trigger for catastrophic collapse in the WAIS.

Were the Antarctic Peninsula to disintegrate, sea levels would rise globally by half a meter (20 inches). The Pine Island and Thwaites Glaciers could add another 1-meter rise. A collapse of the entire WAIS would raise sea levels 5 to 6 meters, perhaps over the course of a century. Combined with the disintegration of Greenland's ice sheet, that could raise the oceans more than 12 meters (40 feet).

If the planet warms enough, it could experience an even greater sea-level rise, since the East Antarctic Ice Sheet is about eight times larger in volume than the WAIS. Some 3 million years ago, when the earth was a little more than 3°C warmer than preindustrial levels (about 2.2°C warmer than today), Antarctica had far less ice and sea levels were a stunning 25 meters (80 feet) higher than today. If we stay on our current emissions path, the planet will almost certainly be that warm by century's end.

"A warming of this magnitude would risk 'the end of civilisation as we know it by the end of this century,'" said Peter Barrett, director of Victoria University's Antarctic Research Centre in Wellington, at the Royal Society of New Zealand's awards dinner in 2004, where he won the New Zealand Association of Scientists' Marsden Medal for lifetime achievement.

2050-2100: THE TRIPLE THREAT

How much the seas rise—and how fast—depends on how hot the planet gets. If we could avoid doubling carbon dioxide concentrations from preindustrial levels, we would have a very good chance of avoiding the worst of sea-level rise and might even avoid melting most of Greenland and Antarctica.

The scenario I put forward in the last chapter assumes that emissions will continue at current growth rates for another decade, then the growth rate slows by half for a decade, and then we aggressively adopt the seven low-carbon technological wedges for five decades (and stop tropical deforestation), and finally emissions start dropping in 2075. Because vicious cycles kick in, this scenario leaves concentrations at more than 800 ppm in 2100 (nearly a tripling of preindustrial levels), with average global temperatures more than 3°C higher than today *and still rising.*

What precisely happens to our coastal cities in a tripled-CO_2 world? You will not find many detailed studies on the subject, for two main reasons. First, most scientists have based their efforts to model climate impacts on a doubling of CO_2 concentrations because they (and their funders) have expected the world to wake up and take action. Second, most climate scientists did not expect the kind of accelerated flow and disintegration of the ice sheets we are now witnessing.

But in a tripled-CO_2 world, the United States should plan on the melting of Greenland and the West Antarctic Ice Sheet (ultimately augmented by ice loss from East Antarctica) to begin fueling a significant sea-level rise this century and beyond. We should plan for a 0.5- to 1.0-meter (20- to 40-inch) sea-level rise by 2050 and a 1.5- to 2.0-meter (60- to 80-inch) sea-level rise by 2100. How likely is this to happen? My interviews with leading climate scientists indicate that these numbers are emerging as the top of the likely range, even for a world of just 700 ppm. You can cut these numbers in half if you are of the cross-your-fingers-and-hope-for-the-best school of disaster planning.

We also need to plan for the probability that, by 2050, the hurricane season we experienced in 2005 will have become fairly typical. Also, the *rate* of sea-level rise by the end of the century is likely to be several inches a decade, and it could be more than a foot a decade. Thus, we are not trying to adapt to a static situation where sea level jumps 3 feet and stops, as so many analysts seem to have assumed. This amount of static sea-level rise has been well studied, but the impact of a constantly rising sea level has not, nor has the synergistic effect of increasing hurricane intensity.

A 1991 study led by the U.S. Environmental Protection Agency (EPA) noted that any significant sea-level rise "would inundate wetlands and lowlands, accelerate coastal erosion, exacerbate coastal flooding, threaten coastal structures, raise water tables, and increase the salinity of rivers, bays, and aquifers." The first 1 meter of sea-level rise inundates about 35,000 square kilometers (13,000 to 14,000 square miles) of U.S. land, roughly half wetlands and half dry land. Many shores would retreat rapidly, with beaches likely to "erode 50–100 meters from the Northeast to Maryland; 200 meters along the Carolinas; 100–1,000 meters along the Florida coast; and 200–400 meters along the California coast."

As sea-level rise increases, the impacts multiply. One 2001 analysis reported that 22,000 square miles of land just on the Atlantic and Gulf coasts are at less than 1.5 meters elevation. While some of that might be salvageable if sea levels rose that high and stopped, in the post-2050 world, steadily rising sea levels would quickly lead to the abandonment of far more area.

Abandonment is particularly likely because the states that have the most area in jeopardy are, in order, Louisiana, Florida, North Carolina, and Texas—Hurricane Alley. By 2100, Louisiana could lose an area the size of Vermont to the sea. Florida could lose a Connecticut. North Carolina and Texas could each lose a Delaware. These numbers are conservative in that they don't consider the impact of tides, storm surges, coastal erosion, or land subsidence. Today, the part of our coast that hasn't been eroded by storm and tide has generally been toughened up by them. Sea-level rise exposes parts of the shore to storms and tides that are not so strengthened.

Now imagine that sea levels keep rising 5 inches a decade or more at the same time that the Gulf of Mexico, Florida, and the South Atlantic coasts are battered year after year by hurricane seasons similar to, or even worse than, what they experienced in 2004 and 2005. There's no chance New Orleans could survive the century. Indeed, the city seems unlikely to survive the next category 4 or 5 hurricane whenever it comes, because so far the U.S. government appears unwilling to foot the bill for designing levees to protect it from such storms—let alone from such storms in a world where sea levels are considerably higher.

If our government won't spend the money to protect New Orleans sufficiently today, what are the chances we will spend the money to protect dozens of coastal cities post-2050, once everyone knows that sea levels will keep rising and intense hurricanes will occur relentlessly? Consider also that by then, we will be devoting huge resources to desperately cutting our greenhouse gas emissions, to figuring out how to reverse catastrophic warming, and to dealing with the devastating consequences of drought, wildfires, and massive relocations. *Protecting* dozens of major coastal cities from future flooding will be challenging enough—*rebuilding* major coastal cities destroyed by super-hurricanes will be an almost impossible task.

This will be the beginning of the era of urban triage. New Orleans, the Outer Banks of North Carolina, the Florida Keys and South Florida, Galveston, and other low-lying Texas cities, Biloxi and other low-lying Gulf Coast cities, will be the first to go. Some will be abandoned before being hit by a category 4 or 5 hurricane. Others, afterward.

In this scenario, most of our coasts, especially along the Gulf and South Atlantic, will be designated permanent (or, more accurately, semipermanent) wetlands and will no doubt be uninsurable for building. Some major ports and cities, such as Houston and Miami, would likely be the subject of major preservation efforts. But I have not seen one estimate of the cost of designing levees and other protections for such large cities against rapidly rising sea levels and a category 5 hurricane.

And this is not the worst case.

The authors of a 2005 study, "Global Estimates of the Impact of a Collapse of the West Antarctic Ice Sheet," led by the University of Southampton in England, point out that theirs is the first paper to consider the global impacts of a 5- to 6-meter (16- to 20-foot) sea-level rise. A 1980 paper by the National Center for Atmospheric Research that considered a similar sea-level rise focused only on the United States. Both these studies are "optimistic" in that they assume that after the ice sheet collapses, the sea-level rise will return to a very slow rate. They also didn't consider how hurricanes will change the cost of any protection measures in Hurricane Alley, or how governments and individuals would perceive the viability of building in those regions.

The 2005 paper's worst-case scenario has some 6 meters of sea-level rise from 2030 to 2130, based on a collapse of the West Antarctic Ice Sheet, with little or no contribution from Greenland. Given recent scientific studies, I believe a more plausible version of the same events might be a steadily accelerating loss of mass from Greenland coupled with periodic collapses of parts of the West Antarctic Ice Sheet, creating much uncertainty and fear.

In their scenario, by 2100, some 400 million people worldwide will be exposed to the rising seas. A total land area of more than 4 million square kilometers will be flooded (absent any protective measures), roughly one-half the area of the continental United States. Low-lying countries such as Bangladesh would be utterly inundated. Trillions of dollars of assets would be at risk. In scenarios where the sea level rises and then stops or slows dramatically (and there is no significant increase in coastal storms), adaptation is straightforward if expensive, and a large fraction of the most populated and valuable coastal lands might well be protected. But if people believed that sea levels would simply continue rising more than a foot a decade, any significant defense of coastal cities would seem untenable, especially in hurricane-prone regions.

In this country, one-quarter of Florida would be submerged—and one-third would be underwater when sea-level rise hit 7.6 meters (25 feet). If one or more category 4 or 5 hurricanes struck what was left of the state on a regular basis, perhaps every other year, how could any significant portion of the state be protected for human population and commerce?

Louisiana would be in the same capsized boat, flooded up to Baton Rouge. In Texas, Galveston, Corpus Christi, Beaumont, and Port Arthur would be submerged. Houston would be seriously at risk and difficult to protect from the combination of rising sea levels and super-hurricanes. Savannah, Georgia, Charleston, South Carolina, Virginia's major coastal cities, one-fourth of Delaware, most land along the Chesapeake Bay, and huge sections of such cities as Washington, D.C., New York, and Boston would be flooded. Large parts of the San Francisco Bay and Puget Sound would also be, although in general the West Coast would be better off since it has fewer low-lying coastal areas and no hurricane risk.

If Americans in 2100 came to see 12 meters (40 feet) sea-level rise as inevitable by 2200, who can even begin to fathom how the nation would respond?

I have focused in this chapter on the "high water" part of the scenario, but let's not overlook the hellish heat we would experience. A November 2005 study in the *Proceedings of the National Academy of Sciences* is one of the few to look at the extreme temperatures that a near tripling of carbon dioxide concentrations would have on United States weather in the last quarter of this century (from 2071 to 2095).

A vast swath of the country would see the average summer temperature rise by a blistering 9°F. Houston and Washington, D.C., would experience temperatures exceeding 98°F for some 60 days a year. Oklahoma would see temperatures above 110°F some 60 to 80 days a year. Much of Arizona would be subjected to temperatures of 105°F or more for 98 days out of the year—14 full weeks. We won't call these heat waves anymore. As the lead author, Noah Diffenbaugh of Purdue University, said to me, "We will call them normal summers." They will be accompanied by extreme droughts on a recurring basis, some in the West lasting for many years at a time, with two to five times the wildfire devastation, as discussed in chapter 3. And temperatures would continue to rise relentlessly into the next century, accompanied by declines in soil moisture over much of this country. Much of the Southwest would be at risk of desertification.

Hell and High Water is not our certain future, but it is the future we should expect and plan for if we do not sharply reverse our energy and environmental policy in the next two decades. As with the avian bird flu, doing nothing would make disaster inevitable.

Scientists once hoped that some as yet unidentified mechanism in the climate system might avert catastrophe, but if climate models have any gaps today, they are gaps that *underestimate* the speed and severity of future impacts.

In any case, even if the worst case of several meters' sea-level rise by 2100 doesn't come to pass, our likely future on a tripled-CO_2 planet Earth is still almost unimaginably grim.

I have left out details of the other impacts scientists see as possible or likely. For example, in a tripled-CO_2 world, the oceans become so warm and acidic that virtually all coral reefs die. In fact, much of the world's oceans, especially in the Southern Hemisphere, become inhospitable to many forms of marine life. Globally, more than a quarter of all species may die, since they are far less capable of adapting than we humans are, especially to such a rapid climate change. In a tripled-CO_2 world, tropical diseases find fertile ground over much larger portions of the planet.

I have focused on impacts in the United States, largely because so many people I talk to mistakenly believe we Americans will not be severely affected by climate change, or at least not anytime soon. But not only will this rich country suffer, poor countries will suffer greatly. Imagine what will happen in Africa, a continent already afflicted with persistent, widespread drought and a shortage of safe drinking water. One 2006 study reported in *Science* found that by 2100, climate change could dry up lakes and streams in one-quarter of the African continent.

Much attention has been given to the possibility that the ocean-circulation patterns could change dramatically, especially by the melting of the Greenland Ice Sheet injecting fresh water into the North Atlantic, which in turn might shut down the so-called thermohaline current that helps warm Europe. This scenario, carried to an absurd extreme in the movie *The Day After Tomorrow*, supposedly plunges the planet into an ice age. This is, as James Hansen put it to me, "the implausible worst-case scenario." While the current may be weakening, a complete collapse is considered very unlikely this century and in any case would be unlikely to put much of a dent in the extreme warming most of the planet is going to experience on our current emissions path.

Some, including James Lovelock, have raised the prospect of a runaway greenhouse effect with ever-accelerating increases in greenhouse gas emissions, which might kill hundreds of millions of people by the end of this century. That view does not find much support in the scientific literature, and I regard it about as implausible as *The Day After Tomorrow.*

The possibility, however, that the thawing tundra might release a large fraction of its carbon in the form of methane is quite genuine, and so we could end up in a quadrupled-CO_2 world (carbon dioxide concentrations of 1,100 ppm) by 2100 or soon thereafter. Such a world is vastly grimmer than anything I have described here.

In a quadrupled-CO_2 world, average temperatures over much of the inland United States would be a scorching 20°F hotter. Soil moisture would drop 50 percent or more over much of the country. Prolonged drought would ravage much of our cropland, turning breadbaskets into dust bowls. Sea-level rise of 80 feet or more would be inevitable. We would exceed global temperatures before the Antarctic ice sheet formed, when sea levels were 70 meters (230 feet) higher on our planet. Humanity would be faced with centuries of suffering from a continuously worsening climate.

Scientists have given us more than enough serious and credible warnings of the consequences of our current path. The IPCC's *Fourth Assessment Report* this year (2007) will present a much stronger consensus and a much clearer and darker picture of our likely future than the *Third Assessment*—but it will almost certainly still underestimate the likely impacts. The *Fifth Assessment,* due around 2013, should include many of the omitted feedbacks, like that of the defrosting tundra, and validate the scenarios described on these pages, especially if we haven't yet sharply reversed our current energy policies. At that point, exceeding a doubling of carbon dioxide concentrations in our air will be a near certainty, and a tripling will be quite likely.

The IPCC process tends to produce an underestimation of worst-case scenarios for two reasons—because it is consensus-based and because it encompasses many greenhouse gas scenarios that assume far stronger action on emissions reduction than the United States or the world seems prepared to embrace.

PART II

THE POLITICS AND THE SOLUTION

CHAPTER FIVE

HOW CLIMATE RHETORIC
TRUMPS CLIMATE REALITY

*The scientific debate is closing (against us) but not
yet closed.*
    —Frank Luntz, conservative strategist, 2002

*Global warming is real (conservatives secretly know
this).*
    —David Brooks, *New York Times* columnist, 2005

The global-warming problem is no longer primarily a scientific
matter. Science has told us what we need to know about how
life on this planet will be ruined if we stay on our current greenhouse gas emissions path. Global warming is also not a technological problem. We have the technologies to avoid the disasters that
await us if we keep doing nothing.

Today, global warming is a problem of politics and political
will. We lack the will to take the necessary actions—and many of
the actions we are poised to take are either inadequate or ill con-
ceived. The great political tragedy of our time is that conservative
leaders in America have chosen to use their superior messaging and
political skills to thwart serious action on global warming, thereby

increasing the chances that catastrophic climate change will become a reality.

Global warming should not be a partisan issue—not when the health, well-being, and security of the next fifty generations of Americans are at stake. But it has become partisan, at least in this country. In order to determine how to create the politics of action in the next decade, we must understand what the politics of inaction has caused in the past decade. That's what this chapter is about.

AMERICA VERSUS THE WORLD

The United States is almost alone in opposing mandatory action to reduce greenhouse gas emissions. The rest of the developed world (other than Australia) believes that the threat posed by warming is so great that they ratified the 1997 Kyoto Protocol, an international climate treaty that requires strong action—reducing emissions to about 5 percent below 1990 levels by 2008–2012. It was a politically difficult move for many of those countries to make given their knowledge that the United States, the world's biggest emitter, would probably not join.

Moreover, the rest of the industrialized world embraced the Kyoto Protocol even though it did not restrict the emissions from developing countries such as China and India, which many in our country see as a fatal flaw in the agreement and a major reason not to vote for it. Yet that flawed agreement is viewed instead in most other countries as a critical first step to solving the climate problem.

British prime minister Tony Blair said in February 2003, "It is clear Kyoto is not radical enough," given the scale of the climate problem. That same year Blair announced that "for Britain, we will agree to the Royal Commission [on Environmental Pollution] target of a 60 percent reduction in emissions by 2050. And I am committed now to putting us on a path over the next few years towards that target," despite the fact that this would force a dramatic change in how England uses energy in transportation, industry, and buildings. In September 2004, Blair gave a speech in which he reiterated Britain's commitment to deep emissions reductions, saying that the accelerating rate of global warming has become "simply unsustainable in the long-term." He went on to explain:

> And by long-term I do not mean centuries ahead. I mean within the lifetime of my children certainly; and possibly within my own. And by unsustainable, I do not mean a phenomenon causing problems of adjustment. I mean a challenge so far-reaching in its impact and irreversible in its destructive power, that it alters radically human existence. . . .
>
> There is no doubt that the time to act is now.

British environmental politics is far, far removed from ours: After the speech, Conservative Party leader Michael Howard accused Blair (the Labor Party leader) of not taking strong enough action and of "squandering the chance to lead efforts against climate change."

The United States has been headed in exactly the opposite direction. President George W. Bush not only rejected the Kyoto Protocol, he has worked feverishly to block other countries from taking any further action to reduce emissions, and he opposes any mandatory action by this country. A major Senate bill from John McCain (Republican) and Joe Lieberman (Democrat) that would put an absolute cap on U.S. greenhouse gas emissions received just 38 votes in the summer of 2005—5 fewer than it received the first time it was offered in 2003—even though its restrictions had been weakened to try to attract support.

"The United States is not going to ratify this process because the U.S. Congress is not going to allow them to do so, even if the

Administration would sign up to it," said John Shanahan, senior counsel to then Senate Committee on Environment and Public Works chair James Inhofe (Republican), in February 2006. Shanahan also predicted, "You need 60 votes in the U.S. Senate to pass anything. They have got 38 right now. And they may go for something 'super light' to win a few more symbolic votes. *But they will never get 60*" (emphasis added).

In 2005, Tony Blair convened a climate conference with dozens of the world's top climate scientists. In 2006 he released a major scientific report, *Avoiding Dangerous Climate Change*, which included more than forty peer-reviewed papers. In his foreword, Blair stated, "It is clear from the [scientific] work presented that the risks of climate change may well be greater than we thought," but he noted, "action now can help avert the worst effects of climate change."

In the United States, climate science is not treated seriously. As many newspaper stories have related, and as a number of scientists confirmed, the U.S. government routinely undermines the ability of government scientists to communicate their ideas to the American public. The administration edits their work and blocks their access to the media. I will return to this point shortly.

In 2006, Fred Barnes, executive editor of *The Weekly Standard*, wrote of Bush's opposition to the Kyoto global-warming treaty:

> Though he didn't say so publicly, Bush is a dissenter on the theory of global warming. . . . He avidly read Michael Crichton's 2004 novel *State of Fear*, whose villain falsifies scientific studies to justify draconian steps to curb global warming. Crichton himself has studied the issue extensively and concluded that global warming is an unproven theory and that the threat is vastly overstated. Early in 2005, political adviser Karl Rove arranged for Crichton to meet with Bush at the White House. They talked for an hour and were in near-total agreement.

Bush ignores every major study by the world's leading climate scientists, ignores his strong ally Tony Blair, yet instead reads Crichton's fiction thriller and spends an hour chatting with him. Apparently, science fiction trumps science fact.

Senator Inhofe praised Crichton for "a compelling presentation of the scientific facts of climate change" and actually invited him to be a witness at a 2005 Senate hearing on the role of science in environmental policy. Crichton took that opportunity to accuse the entire scientific community of fudging the science of climate change, a charge he also makes in his book, and one that meteorologist William Gray made at the same hearing.

THE CONSERVATIVE CONSENSUS ON CLIMATE

Those who deny that global warming is an urgent problem and those who seek to delay strong action to reduce greenhouse gas emissions have been more persuasive than climate scientists. I call these people the Denyers and Delayers, and they have been particularly persuasive among conservatives, who currently hold much of the political power in this country. Let's explore a few examples from the conservative media, blogs, pundits, think tanks, and politicians. I've chosen them to show that the misinformed skepticism about climate science among conservative political and intellectual leaders runs deep and wide.

In February 2006, *New York Times* columnist John Tierney wrote: "Scientists agree that the planet seems to be warming, but their models are so crude that they're unsure about how much it will heat up or how much damage will be done. There's a chance the warming could be mild enough to produce net benefits."

Tierney has packed a great deal of misinformation into two sentences.

The overwhelming majority of scientists agree the planet *is* warming—not "seems to be"—the data itself is beyond dispute.

The models are not crude. On the contrary, the models have become very sophisticated and even predictive. If there is an appearance of uncertainty about how much the planet will heat up or how much damage will be done, it's because of the uncertainty of how much greenhouse gases we humans are going to release into the atmosphere. Scientists spend a great deal of time analyzing and publicly discussing scenarios that include both very low growth and very high growth in human-caused emissions and concentrations. That means scientists talk about a wide range of potential impacts, which may look to some like uncertainty. Tragically, however, the low-emission scenarios have become more unlikely with each passing year of political inaction—inaction driven in large part by influential but misinformed people such as Tierney.

Similarly, it is a dangerous myth that global warming could be mild enough to provide net benefits. This possibility has died because we did not seize the moment, thanks in part to those such as Tierney who have successfully argued for inaction based on the myth itself. For warming to be mild and even beneficial requires first that the climate's sensitivity to forcing by greenhouse gases be on the very low side—a possibility that has been all but eliminated by a stream of studies in recent years. It requires the United States and the world to stabilize atmospheric carbon dioxide *concentrations* at levels low enough to avoid starting the vicious cycles of the carbon system, well below 550 parts per million (ppm). Unfortunately, we are headed to well over 700 ppm. As we saw in the last chapter, exceeding 700 ppm would probably mean another sizzling 3°C or more of warming *this century,* widespread droughts, and an eventual sea-level rise of 40 to 80 feet or more, an outcome neither mild nor beneficial.

Those stark facts mean we must start reducing the amount of U.S. greenhouse gas *emissions* immediately. Yet Tierney opposes "spending large sums to avert biblical punishments that may never come." On Tierney's path of inaction, the only real scientific question becomes, How bad will the impact of global warming be—very serious or irreversibly apocalyptic?

Second, consider the late commentator Jude Wanniski. In May 2005 he wrote an open letter to *The New Yorker,* expressing his disapproval of Elizabeth Kolbert's three-part series "The Climate of Man," which he labeled "Un-Journalism." The only scientific critique he offered was that the series begins by "announcing that the scientific community has now concluded that mankind in a significant way is producing the carbon dioxide that is cooking the atmosphere" and then shows pictures of melting glaciers, "although the reader cannot tell from looking that the glacier is melting because too many of us are driving SUVs or because solar activity in the last part of the 19th century heated up the earth by a degree or two, and the icecaps are still melting as a result."

In fact, scientists have studied solar activity for decades and have concluded that its contribution to recent warming is at most very small. Even stranger, the planet *cooled* slightly in the last part of the nineteenth century, primarily due to multiple volcanic eruptions, including that of Krakatoa, whose particulates helped block out the sun. It's surprising that a smart man like Wanniski would trot out an old and discredited argument—and that he would so easily believe that the entire scientific community had missed this important contribution to the earth's climate. This is just one example of the Denyers ignoring the thousands of studies disputing their position and instead grasping onto notions that have been widely refuted by scientists.

Third and fourth, consider columnists Charles Krauthammer and George Will on the subject of hurricanes and climate change. Krauthammer proclaimed on September 9, 2005, "There is no relationship between global warming and the frequency and intensity of Atlantic hurricanes. Period." He provided no evidence in support of this statement.

On the September 25, 2005, broadcast of ABC's *This Week,*

George Stephanopoulos and David Gergen discussed the recent scientific evidence linking hurricane intensity to global warming. Then Gergen, who is so well known for his political moderation that he has served both Democratic and Republican presidents, said, "It does seem to me under these circumstances this is a wake-up call to take global warming and climate change more seriously." George Will was ready with a sarcastic reply:

> I have an alternative theory. I think these two hurricanes were caused by the prescription drug entitlement. You will say, "How can you say that? The entitlement hasn't even started." There's no conclusive evidence that global warming, that is to say, an unprecedented, irreversible, and radical change has started. You will say, "There's no scientific proof." Same answer. You will say, "Aren't you embarrassed, Mr. Will, to be attaching your political agenda to a national disaster?" Yeah, I'm embarrassed, but everyone else is doing it.

This may well be the most antiscientific statement Will has ever uttered, if not the silliest, equating a serious cause-and-effect relationship put forward by leading scientists using a widely accepted scientific theory with a causation that is sheer nonsense. Once again comes the accusation that anyone who raises this issue has a "political agenda," when clearly those who dismiss it have the agenda.

Strangely, neither Krauthammer nor Will comes from the wing of the conservative movement that reflexively dismisses key scientific theories, such as evolution. Quite the reverse. Within weeks of their global-warming comments, both wrote strong op-eds against those who embrace the "phony theory," as Krauthammer put it, of intelligent design over evolution. Will's reply to school board members who endorsed a proclamation that "evolution is not a fact" was "But it is."

How can such ardent defenders of the science of evolution be such ardent rejecters of the science of global warming? How can people, even those who question the science underlying evolution, embrace the warnings of scientists that avian bird flu could evolve into a powerful human pandemic but reject warnings from climate scientists? How can so many conservatives dismiss the consensus of thousands of the world's top climate scientists?

The answer is that ideology trumps rationality. Most conservatives cannot abide the *solution* to global warming—strong government regulations and a government-led effort to accelerate clean-energy technologies into the market. According to Jude Wanniski, Elizabeth Kolbert's *New Yorker* articles did nothing more "than write a long editorial on behalf of government intervention to stamp out carbon dioxide." His villain is not global warming; it is the threat to Americans from government itself.

George Will's review of Michael Crichton's *State of Fear* says:

> Crichton's subject is today's fear that global warming will cause catastrophic climate change, a belief now so conventional that it seems to require no supporting data. . . .
>
> Various factions have interests—monetary, political, even emotional—in cultivating fears. The fears invariably seem to require *more government subservience to environmentalists* and *more government supervision of our lives.* [Emphasis added.]

Conservatives such as Will are so opposed to government regulations that they are skeptical of anyone who identifies a problem that requires regulatory solutions—and they are inherently accepting of those who downplay such problems.

George Will believes that advocates for action on climate want more government supervision of our lives. But if we hold off on modest government action today, we will almost guarantee the need for much more extreme government action in the post-2025 era. Only Big Government—which conservatives don't want—can relocate millions of citizens, build massive levees, ration crucial resources such as water and arable land, mandate harsh and rapid reductions in certain kinds of energy—all of which will be inevitable necessities if we don't act now.

THE ART OF POLITICAL PERSUASION

> *Of all the talents bestowed upon men, none is so precious as the gift of oratory. He who enjoys it wields a power more durable than that of a great king.*
>
> —Winston Churchill

Anyone who wants to understand the politics of global warming, and anyone who wants to change the politics of global warming, must understand why the Denyers are so persuasive in the public debate and why scientists are not. Science and logic are powerful systematic tools for understanding the world, but they are no match in the public realm for the twenty-five-century-old art of verbal persuasion: rhetoric.

While logic might be described as the art of influencing minds with the facts, rhetoric is the art of influencing both the hearts and minds of listeners with the figures of speech. The figures are the catalog of the different, effective ways that we talk, including alliteration and other forms of repetition, metaphor, irony, and the like. The goal is to sound believable. As Aristotle wrote in *Rhetoric*, "Aptness of language is one thing that makes people believe in the truth of your story."

The figures of speech have been widely studied by marketers and social scientists. They turn out to "constitute basic schemes by which people conceptualize their experience and the external world," as one psychologist put it. We think in figures, and so the figures can be used to change the way we think. That's why political speechwriters use them. To help level the rhetorical playing field in the global-warming debate, I will highlight the three rhetorical elements that are essential to modern political persuasion.

First, simple language. Contrary to popular misconception, rhetoric is not big words; it's small words. "The unreflecting often imagine that the effects of oratory are produced by the use of long words," a precocious twenty-three-year-old Winston Churchill wrote in an unpublished essay on rhetoric. "All the speeches of great English rhetoricians . . . display a uniform preference for short, homely words of common usage." We hear the truth of his advice in the words that linger with us from all of the great speeches: "Judge not that ye be not judged," "To be or not to be," "Lend me your ears," "Four score and seven years ago," "blood, toil, tears and sweat," "I have a dream."

In short, simple words and simple slogans work.

Second, repetition, repetition, repetition. Repetition makes words and phrases stick in the mind. Repetition is so important to rhetoric that there are four dozen figures of speech describing different kinds of repetition. The most elemental figure of repetition is *alliteration* (from the Latin for "repeating the same letter"), as in "compassionate conservative." Repetition, or "staying on message," in modern political parlance, remains the essential rhetorical strategy. As Frank Luntz, a leading conservative-message guru and political strategist, has said, "There's a simple rule: You say it again, and you say it again, and you say it again, and you say it again, and you say it again, and then again and again and again and again, and about the time that you're absolutely sick of saying it is about the time that your target audience has heard it for the first time."

Third, the skillful use of tropes (from the Greek for "turn"), figures that change or turn the meaning of a word away from its literal meaning. The two most important tropes, I believe, are metaphor and irony. "To be a master of metaphor," Aristotle writes in *Poetics,* is "a sign of genius, since a good metaphor implies intuitive perception of the similarity in dissimilars." When Bush said in 2006 that the nation was "addicted to oil," he was speaking metaphorically. Curing an addiction, however, requires far stronger medicine than the president proposed: America could become energy-independent, but only through a series of government-led policies identical to the ones needed to avoid catastrophic climate change.

SCIENCE, CLIMATE, AND RHETORIC

Rhetoric works, and it works because it is systematic. As Churchill wrote, "The subtle art of combining the various elements that separately mean nothing and collectively mean so much in an harmonious proportion is known to very few." Unfortunately, the major player in the climate debate, the scientific community, is not good at persuasive speech. Scientists might even be described as anti-rhetoricians, since they avoid all of its key elements.

Few scientists are known for simple language. As the physicist Mark Bowen writes in *Thin Ice,* his book about glaciologist Lonnie Thompson: "Scientists have an annoying habit of backing off when they're asked to make a plain statement, and climatologists tend to be worse than most."

Most scientists do not like to repeat themselves because it implies that they aren't sure of what they're saying. Scientists like to focus on the things they *don't* know, since that is the cutting edge of scientific research. So they don't keep repeating the things they *do* know, which is one reason the public and the media often don't hear from scientists about the strong areas of consensus on global warming.

Scientific training, at least as I experienced it, emphasizes sticking to facts and speaking literally, as opposed to figuratively or metaphorically. Scientific debates are won by those whose theory best explains the facts, not by those who are the most gifted speakers. This view of science is perhaps best summed up in the motto of the Royal Society of London, one of the world's oldest scientific academies (founded in 1660), *Nullius in verba:* take nobody's word. Words alone are not science.

Scientists who are great public communicators, such as Carl Sagan and Richard Feynman, have grown scarcer as science has become increasingly specialized. Moreover, the media likes the glib and the dramatic, which is the style most scientists deliberately avoid. As Jared Diamond, author of *Collapse,* has written, "Scientists who do communicate effectively with the public often find their colleagues responding with scorn, and even punishing them in ways that affect their careers." After Carl Sagan became famous, he was rejected for membership in the National Academy of Sciences in a special vote. This became widely known, and, Diamond writes, "Every scientist is capable of recognizing the obvious implications for his or her self-interest."

Scientists who have been outspoken about global warming have been repeatedly attacked as having a "political agenda." As one 2006 article explained, "For a scientist whose reputation is largely invested in peer-reviewed publications and the citations thereof, there is little professional payoff for getting involved in debates that mix science and politics."

Not surprisingly, many climate scientists shy away from the public debate. At the same time, the Bush administration has muzzled many climate scientists working for the U.S. government, as we will see. As a result, science journalists, not practicing scientists, are almost always the ones explaining global warming to the public. Unfortunately, the media is cutting back on science reporting in general and finds reporting climate science particularly problematic.

It is not surprising, then, that the American public is so uninformed about global warming, so vulnerable to what might be called the conservative crusade against climate. I say conservative, rather than Republican, because many moderate Republicans have been as strong on climate as Tony Blair, most notably California governor Arnold Schwarzenegger, who said in 2005, "I say the debate is over. We know the science, we see the threat, and the time for action is now." He then committed the state to reduce greenhouse gas emissions to 80 percent below 1990 levels by 2050—precisely the reductions needed to ensure that the Greenland Ice Sheet does not melt. And in 2006, he signed a law crafted with the help of Democratic state legislators that requires a 25 percent reduction in California's carbon dioxide emissions by 2020.

A NOTE ON SKEPTICISM

The people I call global-warming Delayers and Denyers are also called "climate skeptics" or "contrarians." I think those terms are misused here. All scientists are skeptics. Hence the motto "Take nobody's word." Skeptics can be convinced by the facts; Denyers cannot. Skeptics do not continue repeating arguments that have been discredited. Denyers do.

A contrarian is "one who takes a contrary view or action, especially an investor who makes decisions that contradict prevailing wisdom." Contrarians may have a good strategy for making money in the stock market, but how many have a hidden agenda to undermine faith in the stock market itself? Moreover, if the scientific consensus somehow reversed itself, the Denyers wouldn't suddenly reverse themselves. They aren't contrarians.

The Denyers and Delayers, as I use the terms, are those who aggressively embrace one or both parts of a twofold strategy. First, they deny the strong scientific consensus that the climate change we are witnessing is primarily human-caused and likely to have serious negative impacts if we don't reverse our greenhouse gas emissions trends. Second, they work to delay this country from taking any serious action beyond perhaps investing in new technology.

Their beliefs were well articulated by Michael Crichton in a 2006 interview: "If you just look at the science, I, at least, am underwhelmed. This may or may not be a problem, but it is far from the most serious problem. If you want to do something, [limiting emissions] is not what to do. We don't at this moment have good technology to do this, if, in fact, it's necessary to do it."

Such is the road to ruin. Those who advance such a view, including President Bush, deserve a strong label. No doubt many Denyers and Delayers are sincere in their beliefs, but the quotes of Luntz and Brooks suggest that some are not. Sincere or insincere, they spread misinformation or disinformation that threatens the well-being of the next fifty generations of Americans. Denyers and Delayers are also not content merely to dispute the work of climate scientists—they are actively engaged in smearing those scientists' reputations.

THE CONSERVATIVE BATTLE PLAN

The Denyers and Delayers do not just have messaging skills superior to scientists (and environmentalists and most progressive politicians), they also have a brilliant strategy, a poll-tested plan of attack. A 2002 memo from the Luntz Research Companies explains precisely how politicians can sound as if they care about global warming without actually doing anything about it. It focuses in particular on casting doubts about the science. The memo can be found on the web, and anyone who cares about the future of America should read it.

Luntz's team has "spent the last seven years examining how best to communicate complicated ideas and controversial subjects." A big fan of rhetorical devices, Luntz specifically urges conservatives to "use rhetorical questions" whenever discussing the environment.

Like any good rhetorician, Luntz says that "it can be helpful to think of environmental (and other) issues in terms of a 'story.' " His next line is stunning: "A compelling story, even if factually inaccurate, can be more emotionally compelling than the dry recitation of the truth."

Luntz explains, *"The three words Americans are looking for in an environmental policy . . . are 'safer,' 'cleaner' and 'healthier,' "* (emphasis in original throughout). So people who want to seem to care about the environment should use those very words often. He also notes:

> *"Climate change" is less frightening than global warming.* As one focus group participant noted, climate change "sounds like you're going from Pittsburgh to Fort Lauderdale." While global warming has catastrophic connotations attached to it, climate change suggests a more controllable and less emotional challenge.

Focus groups are nothing new in politics, nor is coming up with the best spin for your ideas. But rarely has it been done with such callous disregard for the gravity of a scientific matter.

Luntz's lessons have been taken to heart in more places than you might imagine. An e-mail message sent in July 2005 from NASA headquarters to the Jet Propulsion Laboratory in Pasadena, California, criticized a web presentation that used the phrase "global warming," stating that it is "standard practice" in the agency to use the phrase "climate change." At the insistence of political appointees, the NASA press office had "a general understanding that when something in this field was written about that it was to be described as climate change and not global warming," as one retired press officer put it in 2006.

Interestingly, "climate change" has become for some conservatives, such as Senator Lisa Murkowski, a phrase to describe the obvious changes in climate we are observing in places like Alaska that (in their thinking) may or may not be caused by human activity, whereas "global warming" is reserved for change that is caused by human emissions of greenhouse gases. Like most scientists, I use the terms interchangeably.

Luntz writes, *"The most important principle in any discussion of global warming is your commitment to sound science.* Americans unanimously believe all environmental rules and regulations should be based on sound science and common sense." Luntz did not invent the phrase "sound science"—a good history can be found in Chris Mooney's book, *The Republican War on Science.* Luntz's strong suit is identifying what phrases work and then convincing conservatives to repeat those phrases over and over. "Sound science" works not only because of its alliteration but because it makes the speaker seem to care about science, even when he or she is actually peddling unsound science.

In theory, "sound science" means genuine peer-reviewed and widely corroborated science, as opposed to speculative Soviet-style "politicized science." In the case of global warming, virtually every single piece of peer-reviewed science supports humans as the primary cause, and as we've seen repeatedly the recent literature strongly suggests the impacts will be somewhere between serious and catastrophic if we don't change course soon.

Luntz's central point is the height of cynicism: *"You need to continue to make lack of scientific certainty a primary issue in the debate. . . . The scientific debate is closing (against us) but not yet closed. There is still a window of opportunity to challenge the science."*

This is one of the great tragedies of our times: For Luntz and a large number of conservatives, global warming is strictly a partisan political issue. He acknowledges that the science is moving against his position, but this does not persuade him. He suggests that conservatives muddy the waters, by providing people with information that supports an erroneous view, so that serious action on global warming can be delayed for as long as possible.

Do conservative political and intellectual leaders truly understand that they are on the wrong side of the scientific debate? *New York Times* columnist David Brooks wrote these astonishing words in 2005: "Global warming is real (conservatives secretly know this)." Delay, delay, delay. That is the goal. But we know that with just one more decade of delay, the only way to be sure the Greenland Ice Sheet doesn't melt would be onerous government action.

The Luntz strategy isn't new. One 1969 tobacco-industry memo famously states, "Doubt is our product since it is the best means of competing with the 'body of fact' that exists in the mind of the general public. It is also the means of establishing a controversy." Other, less famous lines are eerily prescient about global warming: "Doubt is also the limit of our 'product.' Unfortunately, we cannot take a position directly opposing the anti-cigarette forces and say that cigarettes are a contributor to good health. No information that we have supports such a claim."

The Denyers and the Delayers are luckier than the cigarette makers because they feel free to tout the "fact" that global warming might have benefits, as John Tierney did in the quote above, or as George Will does when he wrote in December 2004 that the climate models don't tell us "how much warming is dangerous—or perhaps beneficial." This sales pitch—combining doubt with the false hope of potential benefit—is one the tobacco companies could only dream of.

DENY, DENY, DELAY, DELAY

In a box labeled "Language That Works," Luntz recommends lines for Republican speeches that have been repeated endlessly in various forms by the Delayers:

> *"We must not rush to judgment before all the facts are in. We need to ask more questions. We deserve more answers. Until we learn more, we should not commit America to any international document that handcuffs us either now or into the future."*

In science, the facts are never completely in, making this a highly effective rhetorical strategy in any scientific debate. And this line of attack can be used equally well in ten or twenty years, or forever, because *"all* the facts" are never in. If we must wait until the painful reality of mega-droughts and rapid sea-level rise are upon us, the point of no return will have long passed.

Paula Dobriansky, the Bush administration's under secretary of state for global affairs, justified U.S. efforts to block further action on climate change at a December 2004 international conference with these words: "Science tells us that we cannot say with any certainty what constitutes a dangerous level of warming, and therefore what level must be avoided."

Apply this "certainty" test to all public policy, and we would never take any action to avoid any future problem. The Pentagon cannot say with any certainty what constitutes a dangerous level of opposing forces. Epidemiologists cannot say with any certainty what constitutes a dangerous number of birds infected with avian flu. Doctors cannot say with certainty what constitutes a dangerous weight. Does that mean we have no army? No avian flu policy? That a 300-pound patient with health problems shouldn't be put on a weight-loss regimen?

A core element of the White House's climate strategy has been to call for more research into climate change, but here we clearly see the administration saying one thing and doing the opposite. The Government Accountability Office reviewed the administration's research effort and in April 2005 came to the stunning conclusion that the effort was missing a major piece required by law—a plan to assess the impact of global warming on "human health and welfare," agriculture, the environment, energy, and water.

The White House's constant call for more research is nothing but a smokescreen. The Bush team has systematically worked to hold back the results of such research, to censor the information about the real dangers of global warming that its own agencies are supposed to provide to the public. For instance, since the 1990s, the U.S. Global Change Research Program had been working on a "U.S. National Assessment of the Potential Consequences of Climate Variability and Change." The Competitive Enterprise Institute (CEI), a conservative think tank funded in part by ExxonMobil, sued the Bush White House, under the little-known Federal Data Quality Act, to remove this comprehensive peer-reviewed study from circulation, labeling the report "junk science." A Freedom of Information Request revealed in 2003 that the White House had secretly asked CEI to sue it to get the nation's premier climate assessment withdrawn.

In short, the White House conspired with an oil-company-funded think tank to block a major government scientific report that sought to spell out the dangers of climate change to Americans. The failure of our government to warn us of the dangers, to provide our people with a national assessment of the potential consequences of climate change, denies Americans the information they need to make decisions.

The White House heavily edited a 2003 report from the U.S. Environmental Protection Agency, removing several paragraphs that described the dangers posed by rising temperatures, as the *New York Times*, CBS News, and other media outlets reported. It actually removed a reference to key findings of the National Academy of Sciences, a study that the president himself had commissioned. Ultimately every substantial conclusion in the EPA report was gutted. Even the sentence "Climate change has global consequences for human health and the environment" was considered too strong to be left in and it was removed.

The White House actually hired Philip Cooney, a former lobbyist for the American Petroleum Institute, to do its scientific censoring.

Much of what we have learned about the censoring comes from a whistleblower, Rick Piltz, a senior associate from the government office that coordinates federal climate-change programs, who resigned in March 2005. His documents showed that the White House had systematically edited reports by government scientists to make the otherwise strong scientific conclusions and consensus seem doubtful. Two days after Piltz's story broke, Cooney resigned from the White House. Within days, he was hired by ExxonMobil, which has devoted millions of dollars to supporting groups that advance the Denyer and Delayer agenda.

More recently, we have learned the shocking extent of the administration's censorship efforts, thanks to reports in the *New York Times, The New Republic,* and *60 Minutes.* The Bush administration has been engaged for a number of years in muzzling government scientists, according to a number of scientists inside and outside the government. I myself have spoken to many scientists—some of whom are afraid of speaking out publicly—and they confirm this. Rick Piltz has launched a website, www.climatesciencewatch.org, that regularly reports on government censorship of climate research.

Political appointees at NASA put in place a policy to limit media access to James Hansen—director of NASA's Goddard Institute for Space Studies—and all NASA climate scientists. After Hansen reported the NASA data showing that 2005 was the warmest year on record, and after he began giving lectures warning that we have at most a decade to sharply reverse our greenhouse gas emissions trends, NASA's public-affairs staff was ordered to review his forthcoming lectures, journal articles, web postings, and media contacts. Hansen was told he would face "dire consequences" if he continued to speak out about climate change.

After Hansen went public with his charges in early 2006, NASA seems to have changed its public-affairs policy, but the muzzling has continued at other government agencies. Interview requests from the media have been routinely rejected. And at agencies such as the National Oceanic and Atmospheric Administration, those media interviews that are granted can occur only if public-affairs staff monitors the conversation. As Hansen said in February 2006, "On climate, the public has been misinformed and not informed."

As we saw in chapter 2, some NOAA meteorologists have been publicly advocating an untenable scientific position—that recent increases in hurricane intensity have been well correlated with recent increases in sea-surface temperatures, but that the temperature increases have nothing to do with global warming. The NOAA meteorologists who take this position seem to have unfettered access to the press, even though few of them are experts on global warming. On the other hand, we rarely hear from the numerous global-warming experts at NOAA, many of whom disagree with the agency's official position. "Scientists who don't toe the party line are being intimidated from talking to the press," says MIT climatologist Kerry Emanuel. "I think it is a very sad situation. I know quite a few people who are frightened, but they beg me not to use their name."

The man in charge of NOAA is Vice Admiral Conrad Lautenbacher, a Bush appointee with a Ph.D. in applied mathematics and forty years of Navy service. At a December 2003 conference in Milan, for instance, he repeated the standard rhetoric: "I do believe we need more scientific info before we commit to a process like Kyoto." But it isn't clear what "scientific info" would impress him. In

2005 remarks shortly after Katrina hit New Orleans, he said of the connection between hurricane intensity and global warming: "People have hunches, certainly everybody can have a hunch, but the information is not there at this point that would allow you to make that connection. We have no direct link between the number of storms and intensity versus global temperature rise."

Lautenbacher describes the scientific studies that disagree with his view as merely "hunches." He then repeats the argument that the increase in hurricane intensity is just part of a natural cycle, completely unaware that the natural-cycle explanation is itself closer to a hunch than a proven theory, as we've seen. In February 2006, Lautenbacher wrote a letter to NOAA staff stating that "a few recent media reports have (incorrectly) asserted that some NOAA scientists have been discouraged from commenting on the question of whether human-caused global warming may be influencing the number or intensity of hurricanes." In reply, Jerry Mahlman, former director for sixteen years of NOAA's Geophysical Fluid Dynamics Laboratory, wrote:

> Contrary to Dr. Lautenbacher's assertions, I state emphatically that climate scientists within NOAA have indeed recently been systematically prevented from speaking freely. A number of NOAA scientists have directly and openly disagreed with Lautenbacher's statements that deny his direct connection with censorship of climate science.

Mahlman further notes that "the ideologically driven distortion of the truth about the relationship between hurricane intensity increases and warming ocean temperatures has been thoroughly refuted" in the scientific literature.

A great many people and businesses are making major investments and plans based on their understanding of the risk that the Gulf region could get hit by another powerful hurricane. Everyone, from those rebuilding the Gulf Coast and the levees to insurance companies to home owners like my brother, are trying to make plans—plans that involve the lives, the life savings, and the livelihoods of millions of people. They must have good information. They all rely on NOAA for the most objective scientific analysis and projections. Repeating over and over again the scientifically untenable claim that the recent spate of intense hurricanes is just a "natural cycle" with no link to global warming is dangerously misleading. Mahlman noted to me: "What value is there in obscuring the truth or flat-out lying about it?"

The global-warming Denyers and Delayers wish to do far more than just stop the public from learning the truth; they attack the credibility of those who try to tell the facts. The most virulent of them is Senator James Inhofe. In July 2003 he said, "With all of the hysteria, all of the fear, all the phony science, could it be that man-made global warming is the greatest hoax ever perpetrated on the American people? It sure sounds like it." Why would climate scientists pull such a horrible hoax? At his 2005 Senate hearing with Michael Crichton and meteorologist Bill Gray, Inhofe and his witnesses repeated the smear that climate scientists fudge their results in order to satisfy their funders and convince them to hand over more money.

Some of these attacks are very sophisticated and use the best rhetorical tricks. In his 2002 strategy memo, Frank Luntz recommends this attack:

> Scientists can extrapolate all kinds of things from today's data, but that doesn't say anything about tomorrow's world. You can't look back one million years and say that proves that we're heating the globe now hotter than it's ever been. After all, just 20 years ago scientists were worried about the new Ice Age.

Let's look at the worries of scientists 20 years ago. A 1977 report by the National Academy warned that uncontrolled greenhouse gas emissions might raise global temperatures 10°F and sea levels 20 feet. A 1979 academy report warned that "a wait and see policy may mean waiting until it is too late." A 1983 report from the Environmental Protection Agency warned that "substantial increases in global warming may occur sooner than most of us would like to believe," and the result of inaction might be "catastrophic." Twenty years ago, the leading American scientists were worried about global warming.

Michael Crichton repeats this attack in his novel *State of Fear,* in which he has one of his fictional environmentalists say, "In the 1970s all the climate scientists believed an ice age was coming." Snookered, columnist George Will picked this up in his glowing review and then repeated it on the March 26, 2006, edition of ABC TV's *This Week* with George Stephanopoulos. This clever and popular attack tries to make present global-warming fears seem faddish, saying current climate science is nothing more than finger-in-the-wind guessing.

The Denyers insist that climate scientists used to believe in cooling and now they believe in warming. Like all good attacks, this one is built around a partial truth, in this case, a milli-truth, one part in a thousand of the truth. Global warming leveled off between 1940 and 1975. As explained in chapter 2, this was largely a result of dust and aerosols sent by humans (and volcanoes) into the atmosphere, which temporarily overwhelmed the already well-understood warming effect from greenhouse gases. In the 1970s, a few scientists wondered whether the cooling effect from aerosols would be greater than the heating produced from greenhouse gases, and some popular publications ran articles about a new ice age. Most climate scientists were far more worried about the long-term greenhouse gas trends, even in the midst of short-term cooling—and they proved to be right.

The aerosol effect was fully explained in the 1980s and became part of scientific modeling "that is in remarkable agreement with the observations," as Tom Wigley, a leading climatologist with the National Center for Atmospheric Research, wrote in a 2003 letter to the U.S. Senate. Ignoring the science, the Denyers keep repeating the fiction as if it were the latest argument, sounding a bit like flat-earthers but much more dangerous. Senator Inhofe used this smear in his 2005 Senate hearing with Crichton, and George Will wrote, "Thirty years ago the fashionable panic was about global *cooling*," and then he cited a number of quotes that seem to support him. In January 2005 the website realclimate.org debunked the whole notion in a post titled "The Global Cooling Myth." They showed that Will's quotes from scientific magazines are misattributed or taken out of context in a way that nearly reverses their meanings.

Since Inhofe, Crichton, and Will are not scientists, they won't get drummed out of their community for repeating what is factually untrue.

A spring 2003 workshop of top atmospheric scientists in Berlin concluded that the shielding effect of aerosols may be far greater than previously estimated. Nobel laureate Paul Crutzen said, "It looks like the warming today may be only about a quarter of what we would have got without aerosols." This conclusion would suggest the planet may be far more susceptible to warming than previously thought. Crutzen noted that aerosols "are giving us a false sense of security right now." A 2005 study led by researchers at the National Oceanic and Atmospheric Administration concluded, "Natural and anthropogenic aerosols have substantially delayed and lessened the total amount of global ocean warming—and therefore of sea level rise—that would have arisen purely in response to increasing greenhouse gases."

The real irony here is that the aerosol-shielding issue, fully explained, gives the public *greater* reason to act preemptively on climate, not less. The entire record of climate science, rather than being a narrative based on fickle fads, is one of relentless, hard-nosed, continual progression of knowledge, which is characteristic of science, as opposed to politics or propaganda.

TRUTHINESS OR CONSEQUENCES

I believe the most effective piece of propaganda on global warming is Michael Crichton's 2004 novel, *State of Fear*. Everywhere I speak, I am asked questions based on unsubstantiated assertions in his book. More than any other single document published on global warming, the book captures the essence of Frank Luntz's vision: "A compelling story, even if factually inaccurate, can be more emotionally compelling than the dry recitation of the truth." In 2005, Comedy Central's Stephen Colbert introduced the word *truthiness* to describe emotional appeals that sidestep the facts. "Truthiness is what you want the facts to be as opposed to what the facts are," says Colbert. "What feels like the right answer as opposed to what reality will support." He might have coined the term for Crichton.

Although a work of fiction, *State of Fear* has a clear political agenda, as evidenced by Crichton's December 7, 2004, press release:

> STATE OF FEAR raises critical questions about the facts we believe in, without question, on the strength of esteemed experts and the media. Although the story is fiction, Michael Crichton writes from a firm foundation of actual research challenging common assumptions about global warming.

In an appendix titled "Why Politicized Science Is Dangerous," Crichton draws a direct and lengthy analogy between climate science and eugenics and Soviet biology under Lysenko, where all dissent to the party line was crushed and some Soviet geneticists were executed. With no evidence whatsoever, he claims that in climate science, "open and frank discussion of the data, and of the issues, is being suppressed." With this he is using an old trick—accuse your opponent of the same nefarious thing you yourself are doing.

Modern science is by nature open and frank. Any country and any laboratory can conduct any research it wants, and can publish it in one of hundreds of serious journals around the world. The scientific community conducts peer reviews of arguments on their merits—that's the gold standard. Just before the mistake-riddled, global-warming-will-cause-an-ice-age movie *The Day After Tomorrow* came out, the journal *Science* published an article by two environmental scientists that concluded, "In light of the paleoclimate record and our understanding of the contemporary climate system, it is safe to say that global warming will not lead to the onset of a new ice age." I have yet to see a critique of Crichton's book by the global warming Denyers and Delayers, even though it is seriously flawed, as we will see.

Crichton's book deserves a brief review here, since it has become a rallying cry for the Denyers and Delayers. On TV, in interviews, and in talks around the country, Crichton continues to cast doubt on the seriousness and urgency of global warming. He thinks the scientific and environmental communities have fabricated the threat and that efforts to manage the emissions of greenhouse gases are misguided. To make his case, Crichton accuses the scientific community of bad faith, as noted, and he distorts the science. He creates a scientist-hero, Dr. John Kenner, who outdebates the book's environmentalists.

Kenner says that real-life climatologist Jim Hansen manipulated the media in a 1988 congressional hearing, and that he's discredited because "Hansen overestimated [global warming] by three hundred percent." Had Crichton checked primary sources, he would have found Hansen's prediction came very close to being exactly accurate. The smear Crichton now cites was created ten years later, when global warming Denyer Pat Michaels shamefully misrepresented Hansen's testimony. Michaels is a visiting scientist with the Marshall Institute and a senior fellow at the Cato Institute—organizations that receive funds from ExxonMobil to advance the Denyers/Delayers agenda.

A full factual debunking of the book can be found on real climate.org. It's a fascinating tale of how misinformation is spread. Crichton even spreads truthiness in his bibliography, mischaracterizing the landmark 2002 National Research Council report, *Abrupt Climate Change,* as follows: "The text concludes that abrupt climate change might occur sometime in the future, triggered by *mechanisms not yet understood.*" This is simply not true. The report concludes plainly, "Abrupt climate changes were especially common when the climate system *was being forced to change most rapidly.* Thus, greenhouse warming . . . may increase the possibility of large, abrupt, and unwelcome regional or global climatic events" (emphasis added).

Why would Crichton mischaracterize the report in his bibliography? Because one of his main goals in the book is to undermine the case that global warming causes *abrupt* climate change and *extreme* weather events. In his story, a mainstream environmental group is plotting to create extreme weather events that will cause the deaths of thousands of people timed to coincide with a conference on abrupt climate change in order to trick the public into accepting global warming as truth. In a bizarre coincidence, the book's climax has the evil environmentalists carefully plan a seismic tsunami—just weeks before an actual tsunami devastated Southeast Asia.

But the truth is stronger than fiction. Seismic tsunamis are caused by earth tremors. *They are not caused by global warming.* Any climate scientist knows that. This is a stunning blunder by Crichton, calling into question his claim to have any understanding of global warming.

Senator Inhofe, Michaels, and other Denyers have actually accused the environmental community of blaming the Indian Ocean tsunami on global warming. The environmentalists did nothing of the kind. "I am appalled that environmentalists are trying to ride on the backs of 160,000 dead people to push their global-warming agenda without any factual basis," Pat Michaels told the online magazine *Grist* in January 2005. He issued his own press release, saying, "Michael Crichton should sue environmentalists who blame the massive death toll from the Indian Ocean's tragic tsunamis on sea level rise for plagiarism."

In a January 2005 piece titled "The Tsunami Exploiters," columnist James Glassman said that Tony Juniper of Friends of the Earth in Britain had said of the tsunami, "Here again are yet more events in the real world that are consistent with climate change predictions." In fact, Juniper was talking about an increase in 2004 of *other kinds* of natural disasters that may be related to global warming. He had already put out a press release explaining that his remarks were made *before* the tsunami had even hit.

A few environmentalists had pointed out that rising sea levels (caused by global warming) coupled with the decline in natural barriers such as coral reefs (caused at least in part by global warming) had made the area more susceptible to the ravages from a seismic tsunami (caused by earthquakes). They had also pointed out that current climate trends could make future tsunamis even more deadly. Every one of those statements is, unfortunately, true. A *Grist* headline summed up the phony attack with biting rhetoric: "Right-Wingers Exploit Tsunami by Accusing Enviros of Exploiting Tsunami."

The smear about the tsunami is part of a systematic, decade-long effort by the Denyers to change the discourse in the media and the environmental community about the connection between extreme weather events and climate change—and to keep advocates of strong action on the rhetorical defensive. Tragically, their efforts have been all too successful.

THE DEATH OF ENVIRONMENTAL MESSAGING

When a group is so thoroughly beaten rhetorically, its members begin to bicker internally, often self-destructively. In 2004, two environmental strategists, Michael Shellenberger and Ted Nordhaus, released a bombshell essay, "The Death of Environmentalism: Global Warming Politics in a Post-Environmental World," based in part on interviews with twenty-five environmental leaders. Their essay started a virulent debate. Anybody who cares about the environment and global warming should hear both sides.

The original essay is passionately argued but supremely misguided. Interestingly, one of the authors' central arguments concerns rhetoric at its most basic.

> Most environmentalists don't think of "the environment" as a mental category at all—they think of it as a real "thing" to be protected and defended. They think of themselves, *literally,* as representatives and defenders of this thing. Environmentalists do their work as though these are *literal* rather than *figurative* truths. They tend to see language in general as representative rather than constitutive of reality. This is typical of liberals who are, at their core, children of the enlightenment who believe that they arrived at their identity and politics through a rational and considered process. They expect others in politics should do the same and are constantly surprised and disappointed when they don't.
>
> The effect of this orientation is a certain *literal-sclerosis*—the belief that social change happens only when people speak a *literal* "truth to power." *Literal*-sclerosis can be seen in the assumption that to win action on global warming one must talk about global warming instead of, say, the economy, industrial policy, or health care.

Had the authors gone on to make a compelling case that a figurative approach to global warming was superior to a literal approach, these paragraphs might have been a powerful launching point. But ironically, they instead play right into the hands of the political masters of figurative language, the global warming Denyers and Delayers. While figurative language certainly makes for more persuasive messaging—a central point of this chapter—wise public policy, at least in the environmental realm, *must* be based on scientific literalism.

Their thirty-page paper argues three main points:

1. Environmentalists, even after spending "hundreds of millions of dollars" in the previous decade and a half "combating global warming," have "strikingly little to show for it."
2. Environmentalists' efforts to enact policy measures to reduce greenhouse gas emissions through regulation (caps on greenhouse gas emissions and higher fuel-economy standards for cars) have failed and are therefore wrongheaded.
3. Environmentalists are mired in group think and "policy literalism," which makes them unable to see that the true solution to global warming is a visionary technological fix, the New Apollo Project, a proposal to spend $30 billion a year for ten years on clean-energy technologies, developing and deploying renewable energy and hydrogen cars.

The first point is self-evidently true. The authors, however, spend virtually no time trying to analyze *why* the message has failed. They simply assume that the message was wrong. They do not discuss at all the brilliant rhetorical seduction by the Denyers and Delayers. This is like trying to understand why John Kerry lost without examining the Bush team's strategy.

The authors also do not notice that global warming has a key difference compared with previous issues on which the environmental community has been successful—clean air and clean water, for instance. Those issues were dramatically visible (terrible smog in our big cities, Lake Erie catches fire), directly affected people's health at the time, and the solutions, though costly, could be put into place relatively quickly with very visible results. The signs of global warming are less visible (especially since much of the environmental community and media stopped talking about those signs, such as extreme weather, until recently), the major impact is a few decades away, and strong action now will not provide quick visible results. What strong action in the next decade will do—and only strong action can do it—is avoid catastrophic climate change. But that is hardly as sellable—with literal or figurative language—as avoiding tens of thousands of deaths next year by cutting smog.

On the second point, environmentalists have indeed utterly failed to get the United States to put even the mildest cap on greenhouse gas emissions or establish stronger fuel-economy standards. Does the failure to achieve these policies prove they are the wrong policies? Not at all.

The fact that the environmental community is bad at messaging should not be mistaken for proof that its message is bad—particularly in the case of an environmental problem unprecedented in human history and in the face of opponents with vastly superior language intelligence and resources. The industrialized nations, including all of Europe, have made serious commitments to reduce greenhouse gas and are putting into place a cap on carbon dioxide emissions. Those countries all have tougher fuel-economy requirements or much higher gasoline taxes or both than does the United States.

America absolutely needs an aggressive technology strategy similar to the New Apollo Project (minus the push for hydrogen cars). Mandatory reduction targets, such as a cap on carbon dioxide emissions, without aggressive technology programs will slow economic growth. But technology programs without mandatory targets won't solve our climate problem. They are a seductively attractive false hope. That's why the Denyers and Delayers are among the biggest supporters of technology programs without mandatory targets.

CHAPTER SIX

THE TECHNOLOGY TRAP AND THE AMERICAN WAY OF LIFE

There is no doubt that the time to act is now. It is now that timely action can avert disaster. It is now that with foresight and will such action can be taken without disturbing the essence of our way of life, by adjusting behaviour, but not altering it entirely.
—Tony Blair, 2005

It's important not to get distracted by chasing short-term reductions in greenhouse emissions. The real payoff is in long-term technological breakthroughs.
—John H. Marburger III,
president's science adviser, 2006

The mantra of the Delayers is "technology" and "technology breakthroughs." Their technological fix to the greenhouse gas problem is, unsurprisingly, not imminent. It is "long-term." But as we have seen earlier, failing to act in the near term—now—will bring about such drastic conditions that soon our only choice will be to react with extremely onerous government policies.

In 2005, British prime minister Tony Blair described the crucial two-prong strategy we must adopt: "We need to invest on a large scale in existing technologies *and* to stimulate innovation into new low-carbon technologies for deployment in the longer term." Future technology will be able to help preserve our way of life in the long term *if and only if* we have already moved "on a large scale" to technologies that already exist. Over the next few decades, we must rapidly deploy available technologies that stop global carbon dioxide emissions from rising. *Then,* in the second half of this century, we must sharply reduce global greenhouse gas emissions by deploying all the new technologies we have developed.

The time to act is now.

VOLUNTARY WARMING

It is hard to imagine that people will use low-carbon technologies on the vast scale needed until they see a financial return for cutting carbon, and that will not happen until spewing out carbon has a significant financial cost. But for carbon to have a cost, the government must either tax carbon dioxide emissions or create a market that establishes a price for emitting carbon dioxide. This second approach would be similar to the system used to trade emissions of sulfur dioxide under the Clean Air Act administered by the EPA. I prefer the trading system. The Bush administration strongly opposes both.

During the 2000 presidential campaign, George W. Bush promised to regulate greenhouse gas emissions in the electric-utility sector by putting a mandated cap on carbon dioxide emissions that would be modeled on what his father put into place in 1990 regarding sulfur dioxide emissions. This helped blur the distinction between Bush and his opponent, Al Gore, who was well known for advocating action on global warming. Many thought this was a sign that Bush was a moderate on the environment, like his father. Not surprisingly, he has not carried through on this promise, and there have been no regulations of any kind on greenhouse gas emissions during his presidency.

"What will never fly is a mandatory cap on carbon," said James Connaughton in a February 2004 briefing. He is the chair of the White House Council on Environmental Quality and thus is supposed to be one of the administration's *advocates* for the environment. In December 2004 the *Financial Times* reported that U.S. climate negotiators had actually worked "to ensure that future additions to the Kyoto protocol on climate change should avoid committing nations to reducing their carbon dioxide emissions." This must be the first time in U.S. history that a presidential candidate promised a particular environmental remedy and four years later his aides had not only ruled it out but were actively undermining other countries' efforts to adopt it.

Conservative message makers such as Frank Luntz realized that it could be politically dangerous to oppose *any* action on global warming, even if their efforts to obfuscate the climate science were successful. Luntz lays out a clever solution to this conundrum in his 2002 "Straight Talk" memo on climate-change messaging:

> *Technology and innovation are the key in arguments on both sides.* Global warming alarmists use American superiority in technology and innovation quite effectively in responding to accusations that international agreements such as the Kyoto accord could cost the United States billions. Rather than condemning corporate America the way most environmentalists have done in the past, they attack us for lacking faith in our collective ability to meet any economic challenges presented by environmental changes we make. This should be our argument. We need to emphasize how voluntary innovation and experimentation are preferable to bureaucratic or international intervention and regulation.

This pro-technology pitch is quite a reversal for conservatives. In the early 1980s the Reagan administration cut funding for energy efficiency and renewable-energy technology and innovation programs by 70 to 90 percent. The Clinton administration began reversing some of those cuts, but in 1995 the conservative Congress under House Speaker Newt Gingrich refused to fund any increases. In fact, the House of Representatives even pursued legislation that tried to shut down all applied research into low-carbon energy technologies. In April 1996, Deputy Energy Secretary Charles Curtis and I wrote "Mideast Oil Forever," an article for *The Atlantic* explaining "how the congressional attack on energy research is threatening the economy, the environment, and national security."

Ultimately, we were able to stave off the worst of the cuts by demonstrating that the Department of Energy's technology-development efforts had achieved a remarkable payback for the country. My old office at the Department of Energy (DOE) is exceedingly good at developing clean-energy technologies and then getting people to use more efficient versions of existing technology (lighting, motors, heating and cooling). Those energy-efficiency efforts, which cost taxpayers a few hundred million dollars, were verified by the National Academy of Science as having saved businesses and consumers $30 billion in energy costs. But, tragically, while we were able to beat back the most brutal cuts, we did not meet our goal of significantly increasing funding for low-carbon and oil-reducing technologies.

By the time Bush took office, Luntz and other conservative strategists realized that since they opposed genuine action on global warming, they needed a way to sound like they were doing something. The result was the dual strategy of advocating voluntary action and touting new technology.

Luntz counsels conservatives that while the wait-for-new-technology strategy is important, "you will still fall short unless you emphasize the voluntary actions and environmental progress already underway." In February 2002, after a year of sustained criticism from Democrats and others for failing to take any action on global warming, the Bush administration set a voluntary target for the nation to reduce greenhouse gas *intensity* by 18 percent by 2012.

The word *intensity* is often dropped in media coverage, because it is a complex concept that means little to most people. But without the word *intensity*, it sounds like the Bush administration actually made a commitment to *reduce* total U.S. greenhouse gas emissions, rather than to *increase* them, which in fact is what they did. Even with the word *intensity*, U.S. emissions are permitted to increase enormously. *Intensity* here means "the amount per unit of economic activity, as measured by gross domestic product (GDP)." Bush's double-talk committed the nation to reduce greenhouse gas emissions per dollar of GDP by 18 percent over a ten-year period, which by the administration's own calculation would lead to an *increase* in total emissions of 14 percent during that ten-year period—since GDP was projected to rise about 32 percent.

The intensity rhetoric also allowed the administration to say that it was trying to do something when it wasn't. The nation's "greenhouse gas intensity" had been dropping at a faster rate than in the Bush proposal (while absolute emissions kept rising). So the administration was able to generate positive public relations for a commitment that actually allowed greater growth in greenhouse gas emissions than would otherwise have occurred.

Greenhouse gas intensity is a misleading metric because what threatens us is the total amount of greenhouse gases in the atmosphere, not the amount of gases relative to our GDP. Greenhouse gas intensity can drop every year forever, and concentrations will still increase enough to raise sea levels 80 feet.

At negotiations in Montreal in November 2005 to develop a follow-up to the Kyoto Protocol, the chief U.S. negotiator, Harlan Watson, continued the administration's steadfast opposition to mandatory controls. He shamelessly claimed that Bush's strategy had led to genuine environmental progress and had cut emissions from the year 2000 to 2003. But that period includes a recession and 9/11, which severely reduced economic activity and travel-related emissions. Also, Bush did not begin his presidency until 2001 and didn't start his "voluntary" strategy until 2002. Since 2002, U.S. emissions have *risen* at a rate of 1 percent per year.

As compelling as voluntary innovation and experimentation may sound, they simply do not bring about an absolute reduction in emissions, although well-designed efforts funded at high levels *can* slow the growth rate, as discussed below. I know this all too well because for five years in the 1990s I helped develop, oversee, and run the DOE programs aimed at technology development and voluntary greenhouse gas reductions.

In 1992, President George H. W. Bush signed an agreement saying that the United States would adopt policies that would return greenhouse gas emissions to 1990 levels by 2000. The so-called Rio climate treaty came into force in March 1994. In its early days, the Clinton administration thought that an aggressive set of voluntary programs, combined with an energy tax, would stop emissions growth. Personally, I didn't like the energy tax, because energy is not the problem, greenhouse gas emissions are. Congress didn't like the energy tax either and killed it.

After the 1994 midterm elections, the Gingrich Congress began canceling or cutting the funds for most of the voluntary programs. By "voluntary programs" I am referring to efforts that were aimed not at developing new technologies but at accelerating their deployment into the U.S. market. Such market-entry programs involve public education or working with businesses, cities, and states to lower the many barriers to new technology. This key distinction between technology *development* and technology *deployment* may seem mundane, but it is one that will prove critical to whether or not this nation can avoid catastrophic global warming without devastating its economy.

TECHNOLOGY AND THE DELAYERS

"The United States is neither ashamed of its position on Kyoto nor indifferent to the challenges of climate change," then secretary of energy Spencer Abraham said in 2003. "The United States is investing billions of dollars to address these challenges." Following the Luntz script, Abraham continued:

> Either dramatic greenhouse gas reductions will come at the expense of economic growth and improved living standards, or breakthrough energy technologies that change the game entirely will allow us to reduce emissions while, at the same time, we maintain economic growth and improve the world's standards of living.

His Energy Department further reported, "Abraham said no technologies currently exist to significantly cut emissions of gases linked to global warming."

Astonishing double-talk, especially considering that Abraham made it in Berlin to a group of European climate-policy experts, and every single European country had already agreed to dramatic greenhouse gas reductions under Kyoto.

Luntz's memo states that the "scientific breakthroughs" argument works best for the Delayers. He recommends saying that "as a nation, we should be proud. We produce . . . virtually all the world's health and scientific breakthroughs, yet we produce a fraction of the world's pollution." A very large fraction—we Americans produce one-quarter of the world's greenhouse gases, which is presumably more than what he means by "a fraction."

Luntz urges politicians to say, "America has the best scientists, the best engineers, the best researchers, and the best technicians in the world." When Bush launched his hydrogen-car proposal during his 2003 State of the Union address, he said, "With a new national commitment, our scientists and engineers will overcome obstacles to taking these cars from laboratory to showroom, so that the first car driven by a child born today could be powered by hydrogen, and pollution-free."

A hydrogen car available for a child born in 2003 will not be available in time to stop the climate crisis, even if hydrogen cars actually could help reduce greenhouse gas emissions in the 2020s, which they cannot.

Luntz recommends that when supporters of environmental regulations argue, "We can do anything we set our sights on" and "American corporations and industry can meet any challenge," Denyers and Delayers should "immediately agree" but then argue that we don't need "excessive regulation" or an "international treaty with rules and regulations that will handcuff the American economy" (Luntz's favorite metaphor). Republicans, he says, should argue that we can achieve environmental goals with good old American technology alone.

A 2005 Luntz strategy document, "An Energy Policy for the 21st Century," again argues *"Innovation and 21st-century technology should be at the core of your energy policy,"* repeating the word *technology* thirty times. In an April 2005 speech describing his proposed energy policy, Bush repeated the word *technology* more than forty times. This time *Business Week* recognized that Bush was following Luntz's script and noted, "What's most striking about Bush's Apr. 27 speech is how closely it follows the script written by Luntz earlier this year." The article also pointed out "the President's failure to propose any meaningful solutions."

In his 2006 State of the Union address, Bush announced that America was addicted to oil and the solution was a push for breakthrough technologies, especially in advanced batteries for cars, bio-fuels, and renewable energy. He proposed his "Advanced Energy Initiative—a 22-percent increase in clean-energy research." But the 2005 federal budget had actually *cut* energy R&D by 11 percent compared with that of the year before. And three years earlier, in his 2003 address, Bush had said the answer to our energy and environ-mental problems was hydrogen cars, and he *cut* the budget for renewable energy and bioenergy to pay for that unjustifiable pro-gram.

At a February 2006 speech at the National Renewable Energy Laboratory in Colorado, Bush repeated the word *technology* two dozen times. A few reporters noted that two weeks earlier, the lab had laid off a number of people, including top researchers in areas that the president said were now a priority. Bush blamed this on "a budgeting mix-up," saying, "Sometimes, decisions made as the re-sult of the appropriations process, the money may not end up where it was supposed to have gone." A more reasonable explanation: Technology rhetoric is nothing more than rhetoric.

For the Delayers, the technology pitch is win-win-win. It makes them sound like they're doing something, even while global-warming emissions keep rising. The strong pitch for developing new technology leaves the false impression that existing technology cannot solve our problems—the absurd point former energy secre-tary Abraham made in the 2003 Berlin speech. And the Delayers can even reap the rhetorical rewards of touting technology as our solution to global warming without actually spending more money on the key technologies.

The technology mantra seductively plays to the American people's optimism, while stealing the argument from optimists who believe, as I do, that our technology is precisely the reason why we *can* agree to cap greenhouse gas emissions. The pitch has boxed progressive politicians (and scientists and environmentalists) into a corner. Both sides—those who want to delay on global warming and those who want action now—say they advocate technology, but in this narrative only the stick-in-the-mud progressives want onerous rules and regulations. No wonder those pursuing action today have had so much difficulty getting political traction—and no wonder the Delayers repeat their mantra so much.

Like the best seductions, the technology pitch contains a half-truth: We *do* need to invest in technology—but we *must* couple that investment with mandatory emissions-reduction targets or else global-warming pollution will continue its dangerous rise.

It is not just delaying politicians who use the technology trap as a strategy—corporate Delayers love it too. One of the biggest funders of efforts to convince the public that global warming is not occurring has been ExxonMobil. Since the president announced his hydrogen-car initiative, the oil and gas company has also funded significant advertising about its research into hydrogen-related technologies. It also helped fund a $100 million clean-technology research program at Stanford University. In an April 2005 *Washington Post* ad, ExxonMobil proclaimed:

> We're now making the largest ever investment in independent climate and energy research that is specifically designed to look for new breakthrough technologies. The world faces enormous energy challenges. There are no easy answers. It will take straightforward, honest dialogue about the hard truths that confront us all.

Sounds so reasonable, except ExxonMobil has been as much a champion of "honest dialogue" as the Luntz memo is about "Straight Talk." ExxonMobil has pumped more than $8 million into think tanks, media-outreach organizations, and consumer and religious groups that advance the Denyer and Delayer agenda, including the Competitive Enterprise Institute, the Hoover Institution, the Hudson Institute, the George C. Marshall Institute, the Tech

Central Science Foundation, and the Center for the Study of Carbon Dioxide and Global Change, which calls CO_2 emissions "a force for good." Exxon also participated in discussions involving a 1998 fossil fuel industry proposal "to depict global warming theory as a case of bad science."

The leading opponent of fuel-economy standards is General Motors. It has spent millions lobbying Congress to make sure the company is not required to build more fuel-efficient vehicles—cars that competitors like Toyota are selling briskly today because they saw the inevitability of rising oil prices and growing customer concern about the environment. GM is also the leading U.S. car company that advocates hydrogen cars, and it spends millions on ads asserting that these cars are right around the corner—absurdly claiming in April 2005 that we are actually at the "endgame" of GM's hydrogen strategy. What a pity that GM's promises ring hollow, and not just because hydrogen cars are decades away from being a plausible greenhouse gas reduction strategy.

When I was at the U.S. Department of Energy in the 1990s, we partnered with GM, Ford, and Chrysler to speed the introduction of hybrid gasoline-electric cars, since increased fuel efficiency was (and remains) clearly the best hope for cutting vehicle greenhouse gas emissions by the year 2025. This partnership was part of an informal deal between the Clinton administration and the car companies in which we did not pursue fuel-economy standards and in return the car companies promised to develop a triple-efficiency car (80 mpg) by 2004. Ironically, in the mid-1990s, the car companies were actively lobbying to *cut* funding for hydrogen-car development and to shift that money into near-term technologies such as hybrids. Even more ironically, the main result of our government-industry partnership (which had excluded foreign automakers) was to motivate the Japanese car companies to develop and introduce their own hybrids first.

In one of the major blunders in automotive history, GM walked away from hybrids as soon as it could—when the Bush administration came in—after taxpayers had spent $1 billion on the program. The result: Toyota and Honda walked in. *GM, which had had a technological lead in electric drives, let its number one competitor, Toyota, achieve a stunning 7-year head start* in what will likely be this century's primary drivetrain. GM was publicly criticizing the future of hybrid technology as late as January 2004, and finally announced later in that year a halfhearted effort to catch up to Toyota.

Let this history give pause to anybody who promotes a purely technology-based solution to greenhouse gas emissions (and gasoline consumption) in the transportation sector. GM and President Bush have it exactly backward. It's not, as they have argued incessantly, fuel-economy standards that cost American jobs and market share. It's the lack of them. And because the future is one of constrained oil supplies, inevitable oil price shocks, and the urgent need to reduce greenhouse gases in the transportation sector, the car companies that will have the most success are the ones that can deliver a practical, fuel-efficient vehicle, especially efficient dual-fuel vehicles that can run on low-carbon alternatives to petroleum. Toyota and Honda figured this out, but GM insists on fighting the future. As a result, it has been hemorrhaging cash and market share, both of which are being claimed by smarter competitors.

Yes, joint government–auto industry research and development makes sense, and yes, perhaps even a subsidy to support switching automakers' manufacturing base to hybrids is warranted, but *only* together with legislation that sharply tightens fuel-economy standards and caps carbon dioxide emissions.

BREAKING THE BREAKTHROUGH MYTH

What technology breakthroughs in the past three decades have transformed how we use energy today? The answer: There really haven't been any. We use energy today roughly the same way we did

30 years ago. Our cars still run on internal combustion engines that burn gasoline. Alternatives to gasoline such as corn ethanol make up under 3 percent of all U.S. transportation fuels—and corn ethanol is hardly a breakthrough fuel. Fuel economy did double from the mid-1970s to the mid-1980s, *as required by government regulations,* but in the last quarter-century, the average fuel economy of American consumer vehicles has remained flat or even declined slightly.

The single biggest source of electricity generation, by far, is still coal power, just as it was 30 years ago. The vast majority of all power plants still generate heat to make steam turn a turbine, and the average efficiency of our electric power plants is about what it was 30 years ago. We did see the introduction of the highly efficient natural gas combined-cycle turbine, but that was not based on a breakthrough from the past three decades—and constrained natural gas supply in North America severely limited growth in gas-fired power, so the share of U.S. electricity generated by natural gas has grown only modestly in 30 years. Nuclear power was about 10 percent of total U.S. electric power 30 years ago, and now it's about 20 percent. But the nuclear energy "breakthrough" occurred long before the 1970s, and we haven't built a new nuclear power plant in two decades, in large part because that power has been so expensive.

We do have many more home appliances, but they still haven't fundamentally changed *how* we use energy. Interestingly, home energy use per square foot has not changed that much even with all those new electronic gadgets, for two reasons. First, my old office at the DOE developed major advances in key consumer technologies, including refrigeration and lighting. Second, efficiency standards for appliances have made the use of those efficient technologies widespread. From the mid-1970s until today, refrigerator electricity use has dropped a whopping three-quarters. Perhaps that should be called a breakthrough, especially because some of the savings came from remarkable improvements in the guts of the refrigerator from

Oak Ridge National Laboratory. But we still use refrigerators pretty much as we did, so in that sense these breakthroughs didn't change how we use energy.

One of the most widely publicized energy-technology breakthroughs occurred in 1986 when researchers at IBM Zurich Research Laboratory discovered a material that conducted electricity with no resistance at considerably higher temperatures than previous conductors. Over the next few years a series of breakthroughs in these high-temperature superconductors were announced. This technology generated great excitement because it held the promise of superefficient electric motors and loss-free long-distance electric transmission lines. Yet all these years later, you may ask, where are all the high-temperature superconductors? They have had very little impact on either electric motors or power transmissions.

"Typically it has taken 25 years after commercial introduction for a primary energy form to obtain a 1 percent share of the global market" (emphasis added). So noted Royal Dutch/Shell, one of the world's largest oil companies, in its 2001 scenarios for how energy use is likely to evolve over the next five decades. Note that this tiny toehold comes 25 years after *commercial* introduction. The first transition from scientific breakthrough to commercial introduction may itself take decades. Consider fuel-cell cars, which get a lot of hype today. Yet fuel cells were invented in 1839, and more than 165 years later we still don't have a single commercial fuel-cell car. We barely have any viable commercial fuel cells for stationary electric power generation.

I tend to think that Shell's statement is basically true, although I believe we could in some instances speed things up—but only with the kind of aggressive technology-deployment programs and government standards that conservatives do not like. Given that we must dramatically reverse greenhouse gas emissions trends over the next 25 years, we *must* focus on technologies that are either commercial or nearly commercial *today*.

Why don't never-been-seen-before breakthroughs change how we use energy? Why don't breakthrough energy technologies enter the market the way breakthroughs in consumer electronics and telecommunications seem to? If we focus on the two most important sectors for global warming, transportation and electricity generation, the answer is fairly straightforward: The barriers to market entry for new technologies are enormous. The entire electric grid—from power plant to transmission line to your house—represents hundreds of billions of dollars in investment, much of which has long since been paid off. We have coal plants and hydropower plants that are several decades old and still running. This keeps electricity widely available, and much lower in price here than in almost any other industrialized country. And it keeps competing technologies at a permanent disadvantage.

The entire gasoline-fueling delivery infrastructure—refineries, pipelines, gasoline stations, and the like—also represents hundreds of billions of dollars of investment that assures widespread availability, low price, and very tough competition for any potential alternative fuel. A comparable investment has been made in automobile manufacturing plants, a key reason why we have not seen a new American car company successfully launched for a very long time.

Perhaps the best example of a breakthrough that is changing the vehicle market is the nickel metal hydride battery currently being used in virtually every hybrid gasoline-electric car today. The key to making hybrids work is the battery. Research on nickel metal hydrides began in the 1970s. In the early 1980s, a U.S. company, Ovonics, introduced nickel metal hydride batteries into the market for consumer electronics. At the DOE we were interested in hybrids in the mid-1990s because a few years earlier Ovonics had developed a version of the battery for cars under a partnership with the government in the U.S. Advanced Battery Consortium.

Hybrids were introduced into the U.S. car market by the Japanese car companies Toyota and Honda in 1997. Sales began to soar after 2000, thanks to improved engineering, high gasoline prices, and government incentives. Even so, in 2005, 8 years after they were introduced, hybrids were only slightly more than 1 percent of new-car sales in the United States. But here we want to know how long before a breakthrough significantly affects how we use energy or how much energy we use. So the question is, How long before hybrids reduce U.S. gasoline consumption?

Consider first that the average car now lasts for nearly 20 years, making it difficult for any breakthrough technology to have a rapid impact on the market. Second, consider that engine technology has gotten dramatically more efficient in the past two decades, but the average vehicle on the road has not. Why not? The efficiency gains have been offset by increased performance (faster acceleration) and the increased weight of the average car (thanks to the growing popularity of sport-utility vehicles and light trucks).

How soon will hybrids begin reducing U.S. gasoline consumption? The best answer is, "Maybe never." Why should hybrids increase the average efficiency of the U.S. cars and light trucks any more than the steady advances in engine efficiency of the past two decades did? The good news is that hybrid drivetrains provide enough efficiency improvement and their electric motors develop such high acceleration that automakers have used the technology to raise both horsepower and fuel economy simultaneously. But a number of hybrid models have been introduced that achieve only a very modest efficiency gain. Moreover, vehicle efficiency must rise significantly over the next two decades just to keep gasoline consumption—and hence greenhouse gas emissions—constant, simply to make up for the increases that would otherwise come from more and more people buying more and more cars and driving farther and farther.

If we want to reduce U.S. oil consumption and greenhouse gas emissions from cars, the most obvious strategy is the one that we already employed successfully to double the fuel economy of our cars from the mid-1970s to the mid-1980s—tougher government mileage standards. No other strategy has ever worked for this country.

The Denyers and Delayers remain tragically stuck with their "we must wait for new technology" rhetoric. Perhaps the most egregious example of this came in January 2006, after six former EPA administrators—five of them Republican, including EPA chiefs for Nixon, Ford, and Reagan—urged the Bush administration to impose mandatory greenhouse gas emissions controls as a way to address global warming. In response, EPA's administrator, Stephen Johnson, said the administration policy is to pursue voluntary programs and technological innovation, rather than mandates and standards. He then said: *"Are we going to tell people to stop driving their cars, or do we start investing in technology [to cut emissions]? That's the answer, investing in those technologies"* (emphasis added).

This astonishing false choice—invest in technology or force people to stop driving their cars—comes from our country's top person for protecting the environment. Johnson can't seem to grasp that today's *existing* technology was yesterday's new technology. Hybrids were once new; now they aren't. They can substantially reduce U.S. greenhouse gas emissions if government standards require them to do so. Technology is no substitute for standards. Technology is what makes standards practical and affordable.

The Delayers don't believe in technology—they believe only in *new* technology, that is, until it is no longer new. The Bush administration not only opposes significantly higher national mileage standards for cars, it is even opposing in court a law passed by the state of California requiring that car companies use existing technologies to cut carbon dioxide emissions per vehicle by 30 percent. The administration argues that carbon dioxide is not a pollutant California can regulate and that this law illegally preempts federal authority in setting mileage standards for cars.

If the Delayers were truly serious about new technology offering the only possible strategy for dealing with global warming, they would propose a far larger budget to develop it. Yet the Bush administration has never increased the total energy R&D budget for the federal government. And worse, when we take out programs that offer little hope in the first half of this century (such as the hydrogen car program), and we subtract the notorious congressional earmarks that have run rampant since 2000 (which often divert funds from well-designed technology programs to pork-barrel projects), we have seen a substantial decline in money for development of clean-energy climate solutions.

Our bill for *imported* oil alone now exceeds $250 billion a year. In total, Americans spend nearly $1 trillion a year on energy. The global-warming damages this country will sustain could run into the trillions of dollars. The core of any strategy to reduce greenhouse gas emissions and oil consumption is energy efficiency and renewable energy. The R&D budget for those technologies (minus hydrogen and earmarks) is a paltry few hundred million dollars a year and has dropped steadily since 2000. The federal government is spending less than $2 per American per year on the best technologies for avoiding 80-foot sea-level rise.

The scale of the global-warming problem warrants spending equivalent to that of the Manhattan Project or the Apollo program or even the Pentagon's current technology program for developing a missile defense. That would give us an advanced energy-technology program of about $10 to $20 billion per year. One way we know that the Delayer "technology only" strategy is empty rhetoric: The funding levels they suggest cannot deal with the problem—and they block all efforts to increase funding.

ADAPTATION AND GEO-ENGINEERING

Two other technology-based strategies for dealing with global warming—or, rather, not dealing with it—are adaptation and geo-engineering. I haven't written much about how we would adapt to Hell and High Water, for several reasons.

For the foreseeable future, the primary focus of our climate policy today must be avoiding that grim outcome. Also, making adaptation a major focus of U.S. climate policy presupposes a political consensus that climate scientists are correct about current and future impacts. Otherwise, how could politicians agree to spend hundreds of billions of dollars adapting to a large rise in sea levels or an increased number of super-hurricanes or the growing risk of mega-droughts? But if we had such a consensus, then the only moral choice would be to direct the vast majority of our resources to avoiding this catastrophe in the first place.

Many Delayers use the idea of adaptation to argue against action now, to create the false hope that global warming will be of a pace and scale that our children and their children can deal with—which, ironically, would be true only if we ignored their advice and took aggressive mitigation action now. After all, how do you adapt to sea levels rising a foot or more a decade until oceans are 80 feet higher or more? How do you adapt to widespread, ever-worsening global mega-droughts—especially in a world that will need as much water and arable land as possible by midcentury to feed perhaps 9 billion people and grow vast amounts of zero-carbon energy crops?

Of course we should develop drought-resistant crops and new levee technology and better desalinization technology. But for the foreseeable future, avoiding global warming should receive ten to one hundred times the funds of any adaptation effort.

Interestingly, when I was at the Energy Department, we tried to launch an effort aimed at both mitigation *and* adaptation, called "Cool Communities." Most cities have dark surfaces and less vegetation than their surroundings, making them as much as 5°F warmer. Reducing this "heat island" effect would cut greenhouse gas emissions from air-conditioning and offset some of the increase in urban temperatures from global warming—and it would even reduce smog formation.

Cooling a city means planting shade trees for buildings and putting light-colored surfaces on buildings, roads, and parking. The government has a key role to play in research and testing to help identify and develop the best roofing and paving materials, in funding computer models for determining the optimal approach to cooling a city, and in disseminating information. Yet even though Cool Communities was probably the most cost-effective adaptation program ever devised, the Republican Congress killed it because it was part of Clinton's plan to reduce global-warming emissions.

I also don't plan to devote much discussion to how we might geo-engineer our way out. Geo-engineering is "the intentional large-scale manipulation of the global environment" to counteract the effects of global warming. Such a strategy presupposes a political consensus that climate scientists are correct about current and future impacts. How else could politicians agree to spend the vast sums of money needed to, say, put in place thousands of satellites around the earth with mirrors to reflect the sunlight, as some have proposed?

Geo-engineering also presupposes that politicians and scientists and the public share a high degree of certainty about all aspects of climate science. Any human-induced engineering project large enough to affect Earth's climate, such as seeding the upper atmosphere with massive amounts of aerosols, is just as likely to have unintended consequences that make things worse. If we had such certainty and consensus about climate science at any time in the foreseeable future, it would *still* be better to focus the vast majority of our resources on reducing emissions, since that strategy carries far less risk.

"The 'geo-engineering' approaches considered so far appear to be afflicted with some combination of high costs, low leverage, and a high likelihood of serious side effects," concluded John Holdren, director of the Woods Hole Research Center and president of the American Association for the Advancement of Science, in 2006.

Moreover, unlike adaptation, which a country can undertake by itself, geo-engineering is necessarily a planetwide strategy that would certainly require approval and coordination by the United Nations. Yet if the United States has not reversed its energy and climate policy by the 2020s, and joined the world community in an aggressive effort to reduce emissions—if the richest, most polluting nation on earth has refused to devote even 2 percent of its enormous wealth to spare the planet from millennia of misery—we will be a pariah nation. We will hardly be in a position to work with other nations in a desperate gamble to reengineer the planet's climate back to what it was before we engineered it into ruins with our emissions.

One might imagine an internationally sanctioned geo-engineering effort that began with small-scale tests and slowly worked up to planetwide deployment in the second half of this century. If we sharply reverse emissions trends in the next decade, we would minimize both the amount of geo-engineering we might need to do and the speed with which we needed to do it, giving us time to get much smarter and making the effort far less risky. If we hit 500 ppm of carbon dioxide in 2050, however, we will probably be on the verge of crossing a threshold that simply cannot be undone by geo-engineering.

Geo-engineering, like adaptation, might be an important post-2050 strategy, but it seems unlikely to be of much value unless we keep concentrations close to, or, preferably, well below, 550 ppm through 2100. And that requires the aggressive deployment of existing and near-term technology in the electricity and transportation sectors, starting immediately.

CHAPTER SEVEN

THE ELECTRIFYING SOLUTION

This analysis suggests that the United States could reduce its greenhouse gas emissions by between 10 and 40 percent of the 1990 level at very low cost. Some reductions may even be a net savings if the proper policies are implemented.
—U.S. National Academy of Sciences, 1991

What are the winning strategies for avoiding climate catastrophe, for avoiding Hell and High Water? This chapter examines the solutions for the power sector. Amazingly, with the right technology strategy over the next two decades, we could cut U.S. carbon dioxide emissions by two-thirds without increasing the total electric bill of either consumers or businesses.

In previous chapters I have touched on a number of aggressive low-carbon strategies or "wedges" we need to achieve over the next five decades to stabilize concentrations below a doubling. Each wedge ultimately avoids the emission of 1 billion metric tons of carbon a year. These are the ones aimed at reducing emissions from electricity and heavy industry:

1. Launch a massive performance-based efficiency program for homes, commercial buildings, and new construction.

2. Launch a massive effort to boost the efficiency of heavy industry and expand the use of cogeneration (combined heat and power).

3. Capture the CO_2 from 800 new large coal plants and store it underground.

4. Build 1 million large wind turbines (or the equivalent in renewables such as solar power).

5. Build 700 new large nuclear power plants while shutting down no old ones.

The biggest climate threat in the power sector comes from traditional coal plants. That's because coal contains more carbon than any other fossil fuel, and a typical coal plant converts only about one-third of the energy in the coal to electricity. The rest is wasted.

As of 2002, we had nearly 1,000 gigawatts (GW) of coal plants worldwide, which was about 40 percent of total global electricity generation. A typical large coal plant is about one gigawatt, or 1,000 megawatts (MW), in size. By 2030, the world is projected to double that to 2,000 GW of coal electricity.

More than a third of the new coal plants are expected to be built in China, but one in six will be here in the United States. Natural-gas plants had been the preferred new U.S. power plant, in part because they are far more efficient and less polluting than coal plants. But high prices for natural gas have made them much more expensive to operate than coal plants.

The coal plants that will be built from 2005 to 2030 will release as much carbon dioxide as all of the coal burned since the industrial revolution more than two centuries ago. On this emissions trajectory, the world would be emitting 10.5 billion metric tons of carbon (38 billion metric tons of carbon dioxide) in 2030. To stabilize atmospheric carbon dioxide concentrations below a doubling of what they were in preindustrial times, we need to keep *average* annual emissions to only 7 billion metric tons during this century.

So if we build these plants, we need to shut them down within two decades. Considering they represent a capital investment of more than $1 trillion, that doesn't seem likely. The only alternative in 2030 would be to retrofit the plants to capture and store the carbon dioxide they release. But virtually all of the planned coal plants are unsuitable for such retrofits.

CARBON CAPTURE AND STORAGE

Carbon capture and storage (CCS), also called carbon sequestration, is an attractive idea across the political spectrum because it might allow us to continue using a major fossil fuel, coal, but in a way that does not destroy the climate. Everyone from the Natural Resources Defense Council to the Bush administration loves carbon sequestration, although not in quite the same way.

Here's what is involved: To permanently store carbon, to keep it out of our atmosphere forever, the carbon dioxide from all power plants must be removed and stored somewhere forever. The carbon dioxide can be captured either before or after combustion—although capturing it before is far easier and cheaper. Coal can be gasified and the resulting syngas can then be chemically processed to generate hydrogen-rich gas and carbon dioxide. The hydrogen-rich gas can be combusted directly in a combined-cycle power plant. The carbon dioxide can be piped to a sequestration site. The whole process is called integrated gasification combined cycle (IGCC).

IGCC technology costs more than traditional coal plants. The total extra costs for this process, including geological storage in sealed underground sites, are currently quite high, $30 to $80 a ton of carbon dioxide, according to the DOE. As the National Coal Council reported in 2003, "Vendors currently do not have an adequate economic incentive" to pursue the technology because "IGCC may only become broadly competitive with" current coal and natural-gas power plants *under a CO_2-restricted scenario.* Thus, "power companies are not likely to pay the premium to install today's IGCC designs in the absence of clear regulatory direction on the CO_2 issue." Unless we promptly put into place restrictions on CO_2 emissions, carbon sequestration will be pushed much farther into the future. Before carbon capture and storage can become a significant factor, we must have a government policy that puts a cap on emissions.

In February 2003 the DOE announced the billion-dollar, ten-year FutureGen project to design, build, construct, and demonstrate a 275-megawatt prototype plant that would cogenerate electricity and hydrogen and sequester 90 percent of the carbon dioxide. The goal is to "validate the engineering, economic, and environmental viability of advanced coal-based, near-zero emission technologies that by 2020" will produce electricity that is only 10 percent more expensive than current coal-generated electricity.

The administration's strategy is either doubly pointless or doubly cynical, depending on your perspective. First, by the time this technology is ready to commercialize in the early 2020s, the world will have built or begun construction on more than a 1,000 GW of coal plants, using traditional technology that is not designed for CCS. Second, we will still need a mandatory cap on carbon emissions to make FutureGen plants viable because they will be more expensive than traditional plants even in the 2020s. Since the Bush administration opposes a mandatory cap, the whole R&D effort looks like another delaying action.

People in the energy business call it NeverGen.

Sequestration has another problem, one that must be solved if carbon capture and storage is going to be a major contributor to greenhouse gas reductions any time soon: where to put the carbon dioxide. The largest potential physical reservoir is the deep oceans. But ocean sequestration poses serious environmental risks and is unlikely to be viable. After all, the oceans are already storing a large portion of the CO_2 we have poured into the atmosphere. And their ability to store CO_2 is likely to diminish this century (a bad outcome we do not wish to hasten), and the increased acidification of the ocean is already posing a threat to marine life.

Tens of millions of tons of carbon dioxide have already been injected into oil fields to enhance recovery of oil—that's one reason we know CCS works. But using carbon dioxide to increase recovery of oil does not help reduce net greenhouse gas emissions, since the oil itself is ultimately burned, releasing CO_2.

Research is focusing on pumping highly compressed liquid carbon dioxide, called supercritical CO_2, into huge geological formations, such as deep underground aquifers. A 2003 workshop on carbon management by the National Academy of Sciences noted, "Less dense than water, CO_2 will float under the top seal atop the water in an aquifer and could migrate upward if the top seal is not completely impermeable."

What's the problem here? Even tiny leakage rates undermine the environmental value of sequestration. If we are trying to stabilize CO_2 concentrations at twice preindustrial levels, a mere 1 percent annual leakage rate could add $850 billion *per year* to overall costs by 2095, according to an analysis by Pacific Northwest National Laboratory. If we cannot be certain that leakage rates are well below 1 percent, the study concludes, "the private sector will find it increasingly difficult to convince regulators that CO_2 injected into geological formations should be accorded the same accounting as CO_2 that is avoided," meaning that you would not be able to give the same economic value to CO_2 injected underground as to CO_2 that was never generated (because of technologies such as wind or efficiency). The analysis notes, "There is no solid experimental evidence or theoretical framework" for determining likely leakage rates from different geological formations.

The flow of CO_2 *into* the ground from 800 GW of coal plants would equal the current flow of oil *out* of the ground. If we are going to store that huge amount of CO_2 inside deep underground aquifers, exhaustive testing will have to be done. Each potential site will need intensive monitoring to guarantee it can store CO_2 with no leaks. Very sensitive and low-cost in situ monitoring techniques must be developed to provide confidence that leakage rates are exceedingly low. The geologic stability of storage sites—think earthquakes—is especially important because a massive release of carbon dioxide could suffocate a huge number of people if it hit a populated area.

To start sequestering a significant amount of carbon dioxide in the 2020s, we must immediately begin identifying, testing, and certifying sites. This will not be easy; after spending billions of dollars and conducting more than two decades of scientific study, we have identified only one site in this country as a safe, permanent repository for nuclear waste—Yucca Mountain in Nevada—and even in that case, we have been unable to achieve the consensus needed to start storing waste in it.

If sequestration proves feasible on a large scale, there is a glimmer of good news: Analysis suggests carbon capture and storage could eventually eliminate much of U.S. electric-sector coal emissions for between $20 and $40 a ton of carbon dioxide. If we had such a price today—and a major effort to identify and certify storage sites—we *might* see significant sequestration start by 2020. Absent such policies, it will be delayed a decade or more. In the meantime, we must avoid building traditional coal plants. The best strategy for that is certainly energy efficiency.

THE TECHNOLOGY STRATEGY THAT WILL WORK

Our top two priorities in energy policy should be to minimize the need for new coal-fired power and to free up inefficiently used natural gas for high-efficiency power generation. Energy efficiency remains by far the single most cost-effective strategy for achieving these goals, for minimizing carbon dioxide emissions into the air.

Most buildings and factories can cut electricity consumption by more than 25 percent right now with rapid payback (under four years). I have worked with companies from Johnson & Johnson to IBM to Nike who have demonstrated this over and over again. My 1999 book, *Cool Companies,* describes a hundred case studies of companies that have cut their consumption substantially, making a great deal of money in the process and reaping other, unexpected benefits as well. Many companies that have pursued efficiency have found gains in productivity, because better-designed buildings improve office-worker productivity and redesigned industrial processes typically also reduce waste and increase output. So why doesn't every profit-seeking outfit do likewise? There are many reasons why most companies do not do what the best companies do, including inertia and lack of information. Also, I found that companies tend to be far more aggressive about efficiency when there are comprehensive government programs helping them.

We have more than two decades of broad experience with very successful state and federal energy-efficiency programs. In short, we know what works.

Perhaps the most cost-effective federal strategy would simply be to replicate, nationally and globally, California's myriad energy-efficiency programs and standards for homes and commercial buildings. From 1976 to 2005, electricity consumption per capita grew 60 percent in the rest of the nation, while it stayed flat in high-tech, fast-growing California. This astonishing achievement is shown in figure 4, which compares electricity consumption in California (in megawatt-hours per person) with that in the rest of the country since 1960.

How was California able to keep per capita electricity consumption flat for three decades? By adopting an aggressive, performance-based energy-efficiency strategy. By performance-based I mean one targeted toward efforts that deliver the most bang for the buck.

Most of the money came from California utilities. One key reg-

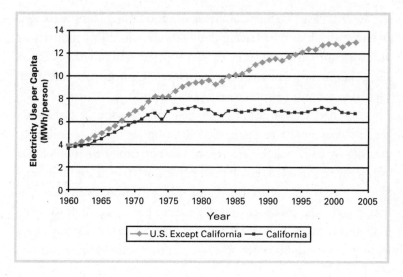

Figure 4. What energy efficiency can really do.

ulatory change was critical. Most utilities in this country can make money only by selling more power, and they lose profits if they sell less. Needless to say, they have little motivation to help their customers cut their electricity bills by using energy more efficiently. California changed the regulations so that utilities' profits are not tied to how much electricity they sell. It has also allowed utilities to take a share of any energy savings they help consumers and businesses achieve. The bottom line is that California utilities can make money when their customers save money.

If it is cheaper to satisfy growing demand with energy efficiency than with new power plants, utilities can still grow their profits. Efficiency strategies today include energy audits, outreach and education, training, technical assistance, and rebates for purchasing energy-efficient products. The California Energy Commission also directly supports efforts to boost energy efficiency, including building codes that specify efficiency requirements for new construction.

You may ask whether California is a good model for the nation, given the troubles it had deregulating its electric-utility industry in the 1990s and the resulting electricity shortages. In fact, the botched deregulation is precisely what convinced Californians that they needed to redouble efforts on energy efficiency.

How had things become botched up? As soon as California began to deregulate in the mid-1990s, utilities cut their efficiency funding in half, causing electricity use per capita to rise. Worse, utilities were forced to sell off their generators, which left them hostage to greedy energy-trading firms such as Enron. Even earlier in that decade, the prospect of deregulation put a near halt to planning and construction of new power plants because nobody knew what the rules and rewards would be in this new deregulated marketplace.

I was at the U.S. Department of Energy at the time, and we warned California that in a deregulated market no one would have an incentive to keep much surplus capacity. Normally, public utility commissions require a lot of spare capacity to ensure that the juice keeps flowing to consumers during the kind of rare long-lasting and widespread heat waves that drive summer air-conditioning demand to extreme levels. Most of the time that spare "peak demand" capacity goes unused, making it relatively unprofitable for companies to maintain. But by the late 1990s, global warming was making those once rare mammoth heat waves commonplace (and in 2006, California would suffer its worst heat wave ever, blanketing the state in 100°F temperature for weeks, killing more than one hundred people and sending electricity demand soaring).

Moreover, California imports a great deal of electricity. In the 1990s, the state failed to anticipate that the rapid growth of neighboring states meant that when the demand crunch came, those imports would dry up. With demand growing faster than expected and supply slowing down, with summers getting hotter and power surpluses shrinking, and with crooked companies like Enron controlling the trade of electricity and natural gas, the day of reckoning was inevitable.

The crisis hit in 1999 and 2000, with shortages and blackouts. The state raised prices and launched a massive efficiency program, the amazing results of which are now in. From 2000 to 2004, California utilities spent $1.4 billion. The average cost of the electricity saved was 2.9 cents per kilowatt-hour—far cheaper than what new peak power generation in the state costs, 16.7 cents per kilowatt-hour, and half the price of building base-load power (generators that run all the time), 5.8 cents per kilowatt-hour. Helping people use electricity more wisely is far cheaper than building new power plants, and that's without even counting the benefits of avoided global-warming pollution and healthier air to breathe.

The utility programs became steadily more effective over time. By 2004, the average cost of the efficiency programs had dropped in half, to under 1.4 cents per kilowatt-hour, cheaper than any form of new power supply in this country—*and far cheaper than any carbon-free power,* including renewable energy and nuclear plants. And it is not just California that has achieved these results. A 2006 report by the Western Governors' Association confirmed that a variety of energy-efficiency programs in western states have delivered savings at similarly low cost.

One of the leaders in California's energy-efficiency push is Dr. Arthur Rosenfeld, the world-class physicist who launched the Center for Building Sciences at Lawrence Berkeley National Laboratory in the 1970s. He helped develop many of the energy-efficiency programs for the state and many of the efficient technologies used around the nation, including windows and lighting. I worked with him at the Department of Energy, and he later became a California energy commissioner, helping guide the state through its crisis.

Rosenfeld told me that California was so satisfied with the efficiency effort that it was going to ramp up funding. Rather than keeping electricity per capita flat, they want to cut it 0.5 percent to 1 percent a year. He notes that the state's efficiency efforts, from the 1970s through 2004, have lowered the energy bill of Californians by $12 billion a year, which comes to a whopping $1,000 a family—even accounting for the extra cost of the efficiency products, services, and programs. The total investment has, on average, paid for itself in energy savings in less than three years and then just keeps generating profits for Californians. *And it is avoiding the emissions of more than 10 million metric tons of carbon dioxide every year.* This is the program to copy—around the country and the world.

I asked Dr. Rosenfeld how much it would cost to duplicate California's program nationwide. His answer: The total effort costs about 2 percent of the revenues of electric utilities, which translates into $6 billion a year nationwide (since the nation's electric bill is about $300 billion). Because parts of United States have more air-conditioning demand than California, he thought the United States might want to invest closer to $9 billion a year, if the goal is to keep electricity consumption per capita flat. Now, that would be a bargain, when America's 100 million households save some $1,000 each year!

Most important, these programs focus on *existing* technology, on getting those technologies into the marketplace, into the homes and offices of consumers and businesses, as quickly as possible. The California Energy Commission has an R&D program to develop new technology, but only so that it can then be sped into the marketplace. In California, efforts to deploy existing energy-efficiency technologies have maintained support through Democratic and Republican administrations alike.

Most conservative politicians do not like energy-efficiency programs, especially ones aimed at accelerating the market entry of new technologies. The Gingrich Congress cut or eliminated most of the deployment programs that the Clinton Energy Department launched in the early 1990s. The Bush administration has sharply cut the funding for the most historically effective efficiency efforts so that they can make room for hydrogen-fuel-cell R&D, which has no realistic prospect of seeing significant marketplace success for several decades.

The Bush administration has relentlessly cut funding for technology deployment. Especially counterproductive is the administration's move to shut down the DOE's regional offices in Boston, Philadelphia, Atlanta, Chicago, Denver, and Seattle. These were set up in the 1970s to help the nation learn how to save energy. They are, or were, the primary national effort to deliver technical and financial assistance to communities, states, industries, and other energy users. The president is shutting them down just when we need them the most.

NATURAL-GAS EFFICIENCY

Since 2000, the United States has suffered through repeated price spikes for natural gas. Between 1999 and 2002 we added some 138 gigawatts in natural-gas-fired capacity, but the increased demand for gas—no surprise—led to a price increase. North American natural-gas supply is limited. Worse, Canadian natural-gas exports to the United States are projected to decline in coming years as Canada uses more and more of the gas for its own purposes, including producing oil from the Alberta tar sands.

High natural-gas prices have driven increases in demand for coal electricity and for new coal plants. High prices have led many politicians to advocate spending tens of billions of dollars on facilities to bring in liquefied natural gas (LNG), even though LNG tankers and terminals are widely seen as a major terrorist target and even though that would increase the nation's dependence on imported energy. About 58 percent of the world's natural-gas reserves are in Iran, Russia, and Qatar—hardly bastions of democracy or stability, hardly the kind of countries we want to be beholden to.

Rather than a major effort to increase our dependence on

foreign-energy supplies, a superior national strategy is more efficient use of our domestic natural gas. As with electricity, most buildings and factories can cut natural-gas consumption by more than 25 percent right now with rapid payback (under four years), after which the savings become profits forever.

A major focus should be on more efficient use of steam, which is crucial for production in energy-intensive industries such as chemicals, food products, plastics, primary metals, pulp and paper, textiles, and petroleum refining. It is generated mainly by natural gas. Steam accounts for $24 billion a year of U.S. manufacturing energy costs and 40 percent of U.S. industrial carbon dioxide emissions. Expanding state and federal efforts to use steam far more efficiently, such as the DOE's Best Practices Steam Program, would cut those numbers sharply.

The energy-intensive industries are not only major consumers of natural gas, they account for 80 percent of energy consumed by U.S. manufacturers and 90 percent of the hazardous waste. They represent the best chance for increasing efficiency while cutting pollution. Many are major emitters of greenhouse gases other than carbon dioxide. A 1993 analysis for the DOE found that a 10 to 20 percent reduction in waste by American industry would generate a cumulative increase of $1.94 trillion in the gross domestic product from 1996 to 2010. By 2010 the improvements would be generating 2 million new jobs, or roughly 1.5 percent of employment in that year.

For these reasons, in the 1990s, the Energy Department began forming partnerships with energy-intensive industries to develop clean technologies. We worked with scientists and engineers to identify areas of joint research into technologies that would simultaneously save energy, reduce pollution, and increase productivity. The Bush administration has slashed funding for this program by 50 percent—and wants to shut it down entirely. This is especially baffling from an administration that opposes environmental regulations, because funding for pollution prevention technology is by far the best way for the nation to minimize the need for such regulations. But that's why it's always important to remember that the new-technology-is-the-only-answer pitch is just empty rhetoric, no matter how many times the administration repeats it.

An important companion strategy to natural-gas efficiency would be a major national effort to encourage the simultaneous generation of both electricity and heat, called *cogeneration,* or combined heat and power. Cogen provides large opportunities to save both energy and carbon dioxide. Right now, fossil fuels burned at large central-station power plants generate most of the electricity used by U.S. companies. These plants are typically quite inefficient, converting *only about one-third* of the energy in fossil fuels into electricity. The waste heat generated by that combustion is literally thrown away, and then more energy is lost transmitting the electricity from the power plant to the factory or building. The total energy wasted by U.S. power generators each year equals the total energy Japan uses each year. More fossil fuels are then burned in our buildings and factories to provide heat, hot water, and steam. The average building boiler converts *only about two-thirds* of its fossil fuels to useful heat or steam.

By generating electricity and capturing the waste heat in a cogeneration system, much energy and pollution can be saved. *Overall system efficiencies can exceed 80 percent.* Total greenhouse gas emissions can be cut in half.

Many studies have shown that the potential market for cogen is enormous. For instance, a 2000 study for the DOE found that the market potential for combined heat and power at commercial and institutional facilities alone was 75,000 megawatts, about one-tenth of current U.S. power-generation capacity. The remaining potential in the industrial sector is about 88,000 megawatts.

Cogen and other *on-site* power systems, such as solar panels, are called distributed energy as opposed to large central-station power plants, like coal or nuclear. Their market penetration is limited by barriers that have nothing to do with their cost or performance—especially the countless obstacles and fees that major utilities can place in their way. In the late 1990s, the DOE launched a study of these barriers, looking at sixty-five distributed-energy projects. The result was a July 2000 report that offered a variety of recommendations we should embrace:

- Adopt uniform standards for interconnecting distributed power to the grid.
- Adopt testing and certification procedures for interconnection equipment.
- Accelerate development of distributed power-control technology and systems.
- Develop tools for utilities to assess the value and impact of distributed power.
- Develop new regulatory principles compatible with distributed-power choices.
- Adopt regulatory tariffs and utility incentives to fit a distributed-power model.

The strategies we need to avoid climate catastrophe are not about imposing the heavy hand of government on the marketplace; rather they are about leveling the playing field and giving an extra push to low-carbon technologies. How much carbon dioxide could an efficiency and cogen strategy save the country? Before answering that, let's look at the potential for renewable power.

THE RENEWABLES REVOLUTION

Energy efficiency can stop the runaway growth of electricity demand. Cogeneration can reduce the carbon emissions of much of the electricity that is generated. Renewable energy can deliver electricity without any carbon emissions. In terms of annual percentage growth, wind and solar energy were the two fastest-growing forms of power in the past two decades. I will focus here on wind because it is the renewable that can meet the most large-scale demand at the lowest price.

Modern wind turbines convert the kinetic energy of the wind into electricity. Wind turbines are often grouped together into "farms" to generate bulk electrical power. Electricity from these turbines is fed into the local utility grid and distributed to customers.

America has exceptional wind resources, especially the central United States from the Texas Panhandle up through the Great Plains. North Dakota alone has enough energy from high-wind resources to supply 36 percent of the electricity of the lower forty-eight states. Much of the available wind, however, is not located near the consumer. Therefore, if wind were to become a significant portion of the generation mix, additional investments in transmission and distribution infrastructure would be needed.

Over the past 15 years, major aerodynamic improvements in blade design have cut the cost of electricity from wind power by 10 percent *per year.* New, utility-scale wind projects are being built all around the country today, delivering electricity at prices as low as 4 cents per kilowatt-hour in the best wind sites. Media attention has been focused on the few public disputes over wind-farm locations, such as the offshore wind farm planned near Cape Cod, but most of the country has been embracing wind enthusiastically; aggregated installed wind in the United States is roughly 9,000 megawatts as of the end of 2005—five times the installed capacity of 1999.

The next-generation wind turbine is projected to bring costs down to 3 cents per kilowatt-hour over the next several years (including the wind-production tax credit). Since wind is an intermittent electricity generator and does not provide power on an as-needed basis, it loses some value on a per-kilowatt-hour basis, compared with traditional generation that can provide steady base-load power. On the other hand, wind can more than make up for this lost value by providing benefits in terms of reduced emissions and elimination of fuel risk (such as seen by natural-gas plants).

While wind now provides less than 1 percent of U.S. electricity generation, it represents up to 40 percent of electricity in regions of Germany, Spain, and Denmark. And wind is only one of several renewable technologies that are near-competitive with grid electricity. As a major 2004 report by the International Energy Agency concluded: "Under the best conditions—optimized system design, site and resource availability—electricity from biomass, small hydropower, wind and geothermal plants can produce electricity at costs ranging from 2–5 cents/kilowatt-hour."

Note that geothermal energy made that list. Geothermal power converts the earth's own deep energy into heat and electricity. It remains a very attractive power source. But in 2006, when Bush proposed his "Advanced Energy Initiative—a 22-percent increase in clean-energy research," he needed to find money to fund it. He found it by zeroing out all federal funding for geothermal research.

Renewable-energy power plants typically have high capital costs, but their operating costs are low, because they don't consume fuel on a daily basis. While most forms of renewable energy are not competitive with current wholesale electricity prices, it is well to remember that

1. Many traditional power plants have long since paid off their capital costs, so that their electricity cost comes only from fuel and operating costs. New fossil fuel power plants don't have that price advantage.
2. Many renewables have not yet achieved their ultimate cost reduction from either improvements in technology or manufacturing economies of scale at higher volume.

3. Carbon dioxide emissions have no economic cost to the producer and are never counted in the comparisons of true energy costs.

So new renewables will be increasingly competitive with new fossil fuel plants, especially when we properly account for the real cost of global-warming emissions, which, as we have seen, threatens to bring about almost incalculable damage to the next fifty generations of Americans.

It is then no surprise to learn that by 2005, some two dozen states and more than forty countries had a national target for their own renewable-energy supply, including all twenty-five countries in the European Union. The E.U. has set a target of having 21 percent of its electricity come from renewables by 2010.

Our Congress, however, refuses to adopt a renewable standard that would require even 10 percent of U.S. power to be delivered by renewable energy. Yet a standard requiring *20 percent* of U.S. electricity to be renewable by 2020 has very little net cost to the country, but brings the huge benefit of reducing future natural-gas prices and future greenhouse gas emissions. Under such a standard, electricity prices would be *lower* in 2020 than they are today, according to a 2001 Department of Energy study.

POWER SWITCH

What could this country achieve with an energy policy based on existing technology and the most successful strategies used by states and other countries to get those technologies into the marketplace? In 2003 I coauthored a study on "The Path to Carbon-Dioxide-Free Power," which focused on the three technology areas I have been discussing: energy efficiency, cogeneration, and renewables.

The results were very promising. They showed that with a set of innovative and ambitious policies the U.S. electricity sector could cut CO_2 emissions in half by 2020. The price of carbon dioxide never exceeds about $15 a ton ($55 a ton of carbon), which translates into slightly more than 1.5 cents per kilowatt-hour added to the cost of a traditional coal plant. Electricity rates rise slightly, but at the same time we will be using electricity more efficiently, which will cause bills to drop substantially. The net savings would be about *$20 billion per year* from 2004 to 2020 and would exceed $80 billion a year after 2020.

The country would see only a small increase in electricity generation from current levels, and natural-gas use stays roughly flat, even while the U.S. population rises 20 percent and industrial output increases 75 percent. Yet consumers and businesses would receive the same or better energy services in 2020 than in the business-as-usual case. The power system would become more reliable and less vulnerable to external disruption, including terrorist attack.

As an added benefit, Americans would see a sharp decline in air pollution and a resulting improvement in health. Utility mercury emissions, which threaten the health and developmental ability of children, would drop 90 percent. Emissions of oxides of nitrogen and sulfur dioxide, which are linked to respiratory and other health problems in humans, drop by two-thirds or more.

Many studies before and since this one have shown similar results, including two major studies by our national laboratories. In fact, *Policy Implications of Greenhouse Warming*, a 1991 study by the National Academy of Sciences, concluded, "This analysis suggests that the United States could reduce its greenhouse gas emissions by between 10 and 40 percent of the 1990 level and at very low cost. Some reductions may even be a net savings if the proper policies are implemented."

American conservatives (and many economists) do not believe such emissions reductions are possible without a very high cost. They do not accept that the economy now operates inefficiently, nor do they believe government technology policies would have much value. Their studies typically ignore the possibility of efficiency and cogeneration, and, with tunnel vision, they assume the only way to achieve the proposed reductions is with very high prices for carbon dioxide in electricity.

If we do nothing for the next two decades, U.S. carbon dioxide emissions will rise another 20 percent or more, and we will have invested hundreds of billions of dollars in another generation of inefficient and carbon-intensive technologies and power plants. If we then try to reduce emissions to 60 percent below 1990 levels by 2050, the cost of energy would probably have to double and the government would probably have to simply mandate shutting down most of our coal plants, with devastating consequences for consumers and businesses. The Delayers believe that action on global warming will hurt the U.S. economy and require onerous government mandates. Ironically, their way of thinking could become a self-fulfilling prophecy.

Having worked with dozens of companies to design profitable emissions-reduction strategies, and having carefully reviewed more than a hundred specific case studies of buildings and factories that employed energy efficiency, cogeneration, and renewable energy, I have no doubt that the United States could dramatically reduce its carbon emissions per kilowatt-hour without raising its overall energy bill.

But you don't have to take my word for this. Nor do you have to wade into the dull details of either the technologies or the economics. Just consider California. In 2004 the state consumed about 7,000 kilowatt-hours per person, whereas the rest of the country consumed about 13,000 kWh per person. California's electricity rates (cents per kWh) are about 50 percent higher than the national average, yet its annual electric bill per person is about the same as the rest of the nation because it wastes less electricity. Its rates are higher partly because California is paying for the legacy of its flawed deregulation in the 1990s, and that portion of the extra rate should decrease over time. Its rates are also higher because it has much cleaner power generation, using more renewables and natural gas than the rest of the country. Californians decided that they value the reduction of unhealthful air pollution.

The result is that each kilowatt-hour consumed in California generates only about half the carbon dioxide emissions of the national average. Combine that fact with the more efficient use of electricity, and you get a startling statistic. In terms of electricity consumption, *the average California generates under one-third of the carbon dioxide emissions of the average American while paying the same annual bill.*

NUCLEAR POWER

The lack of knowledge about energy by even the most senior politicians is scary. Consider Senator John McCain's comments in a March 2006 interview, in which he stated he would demand legislation to expand U.S. nuclear power as part of his efforts to reduce greenhouse gas emissions: "It's the only technology presently available to quickly step up to meet our energy needs," he said.

Wrong on both counts: Nuclear is not the only technology, nor is it the quickest. The licensing and construction process for nuclear plants takes many years, and it should, given that the plants are expensive, carry many safety and environmental risks, and have been given limited liability by Congress in case of an accident. An energy-efficiency strategy would be much faster.

McCain's comment reflects a common misconception that some never-named entity is mysteriously holding back the expansion of nuclear power in this country. What has really been holding back nuclear power is the economic and other risks it poses to utilities and financiers—Wall Street.

Nuclear energy is mostly carbon-free power. Yet it is not a slam-dunk solution to global warming. A major 2003 study by MIT, "The Future of Nuclear Power," highlighted many of the "unresolved problems" that have created "limited prospects for nuclear power today." The study found that "in deregulated markets, nuclear power is not now cost competitive with coal and natural gas." The public has significant concerns about safety, environmental, health, and terrorism risks associated with nuclear power. The study also found that "nuclear power has unresolved challenges in long-term management of radioactive wastes." It described possible technological and other strategies for addressing these issues but noted that "the cost improvements we project are plausible but unproven."

Peter Bradford, a former member of the Nuclear Regulatory Commission (NRC), told the *New York Times* in May 2005, "The abiding lesson that Three Mile Island taught Wall Street was that a group of NRC-licensed reactor operators, as good as any others, could turn a $2 billion asset into a $1 billion cleanup job in about 90 minutes."

Nuclear power may well be one important piece of the climate-stabilization puzzle, which is why I have included it as one of the eight wedges. Achieving one nuclear wedge means building a nuclear power plant somewhere in the world every month for the next fifty years, while maintaining current nuclear capacity.

But nuclear power is hardly a fledgling technology that needs even more targeted support from the U.S. government. Nuclear already gets countless subsidies. For instance, the Price-Anderson Act limits liability in the event of a nuclear disaster. And the Energy Policy Act of 2005 gives the industry billions of dollars more in subsidies—even authorizing more than $1 billion to build a nuclear plant solely for the purpose of making hydrogen, an especially pointless subsidy, as we will see in the next chapter.

The nation needs to put into place mandatory carbon dioxide controls. If a significant price for carbon makes nuclear power attractive to utilities and financiers, and if the plants meet the necessary safety and environmental codes, and if the country can finally agree on a place to put the nuclear waste, then new nuclear plants may well make a significant contribution to reducing greenhouse gas emissions in this country. I certainly wouldn't shut down any existing nuclear plants that are run safely. Nor would I discourage other countries from pursuing nuclear power, as long as it is done under the proper international controls to prevent the proliferation of nuclear weapons.

California, however, achieves its remarkably low per capita carbon dioxide emissions from electricity while getting a lower share of its power from nuclear energy than the national average. That's why federal electricity policy should focus on establishing a price for carbon dioxide, promoting energy efficiency, cogeneration, and renewable energy, and accelerating coal gasification together with carbon capture and storage. Those strategies can take us as far as we need to go on emissions reductions in the utility sector for the next few decades.

We will need a similarly aggressive and intelligent set of technology policies to deal with the other major CO_2-producing sector of the U.S. economy—transportation.

CHAPTER EIGHT

PEAK OIL, ENERGY SECURITY, AND THE CAR OF THE FUTURE

We have a serious problem. America is addicted to oil, which is often imported from unstable parts of the world.

—President George W. Bush, 2006

In the absence of revolutionary changes in energy policy, we are risking multiple disasters for our country that will constrain living standards, undermine our foreign-policy goals, and leave us highly vulnerable to the machinations of rogue states.

—Senator Richard Lugar, 2006

Our ever-worsening addiction to oil makes America less secure. Since 1990, we have fought two wars in the Persian Gulf. We suffered a major terrorist attack funded largely by Persian Gulf oil money. Every year we send more than $250 billion overseas because we import most of our oil. Oil prices keep spiking above $70 a barrel, and gasoline above $3 a gallon. The economic lifeblood of our country is held hostage to countries that are antidemocratic and politically unstable—and to terrorists who keep targeting the world's oil infrastructure. Price spikes above $100 a barrel (and $4 a gallon) are all but inevitable in the coming years. And many fear we may be close to seeing worldwide oil production peak and then decline, which will bring an era of steadily rising oil and gasoline prices.

It's no wonder that politicians—even those who don't worry about global warming—keep talking about oil. So why haven't we taken any serious action on oil for decades? The answer is simple—reducing U.S. oil consumption requires a major government-led effort, such as much tougher mileage standards, and our political leaders have rejected such efforts (except for ones that are merely cosmetic).

The astonishing January 2006 statement by President Bush's EPA administrator, Stephen Johnson, bears repeating: "Are we going to tell people to stop driving their cars, or do we start investing in technology? That's the answer, investing in those technologies." This false choice leaves the nation with no oil policy except strong, empty rhetoric suggesting that the cure for our addiction to oil can be found in happy talk about future technology. Here's what President Bush said the next month, in February 2006:

> Our nation is on the threshold of new energy technology that I think will startle the American people. We're on the edge of some amazing breakthroughs—breakthroughs all aimed at enhancing our national security and our economic security and the quality of life of the folks who live here in the United States.

The president has actually misdirected more than a billion dollars toward the development of hydrogen cars, a solution that will not address either our oil or climate problems in our lifetime, as we will see. I also examine in this chapter why the peak in global oil production is less of a threat to our way of life than is widely perceived, and why peak oil won't avert catastrophic climate change.

We will see why the win-win policies needed to avoid Hell and High Water would also make this nation energy-independent by midcentury, even with declining domestic oil supplies. Finally, this chapter describes the car and fuel of the future. Let's start with some background.

TRANSPORTATION AND OIL

About two-thirds of U.S. oil consumption is in the transportation sector, the only sector of the U.S. economy almost wholly reliant on oil. The energy price shocks of the 1970s helped spur growth in natural-gas use for home heating. It also drove the electric utility sector and the industrial sector to sharply reduce their dependence on petroleum. But roughly 97 percent of all energy consumed by our cars, sport-utility vehicles, vans, trucks, and airplanes is still petroleum-based.

Over the past two decades, cleaner engines and reformulated gasoline have worked together to cut vehicular emissions of noxious urban air pollutants, especially the oxides of nitrogen that are a precursor to ozone smog and particulates, the stuff that does so much damage to our hearts and lungs. But the contribution to global warming by cars and light trucks, such as SUVs, has risen steadily. In the 1990s the transportation sector saw the fastest growth in carbon dioxide emissions of any major sector of the U.S. economy. And the transportation sector will generate nearly half of the 40 percent rise in U.S. carbon dioxide emissions forecast for 2025.

Internationally, the situation is equally fearful. As Claude Mandil, executive director of the International Energy Agency (IEA), said in May 2004, "In the absence of strong government policies, we project that the worldwide use of oil in transport will nearly double between 2000 and 2030, leading to a similar increase in greenhouse gas emissions." If by 2050 the per capita energy consumption of China and India were to approach that of South Korea, and if the Chinese and Indian populations increase as predicted, those two supergiant countries *by themselves* would consume more oil than the entire world used in 2003.

"It took us 125 years to use the first trillion barrels of oil," says a Chevron oil-company ad. "We'll use the next trillion in 30." This computes to an average of about 33 billion barrels of oil a year, which is 91 million barrels of oil a day or *A Thousand Barrels a Second*, the title of a 2006 book by energy economist Peter Tertzakian. World demand hit about 84 million barrels a day in 2005, up from 78 million in 2002—a torrid pace of demand growth that slowed only somewhat when prices spiked above $60 and even $70 a barrel in 2005 and 2006.

Since oil is a finite, nonrenewable resource, many analysts have tried to predict when global production will peak and then start declining, just as U.S. oil production in the lower forty-eight states did three decades ago. Some experts believe this peak will occur by 2010. Princeton geophysicist Kenneth Deffeyes wrote in 2001, "There is nothing plausible that could postpone the peak until 2009. Get used to it." Royal Dutch/Shell, a company that has downgraded its own oil-reserve estimates, adds only two or three decades to this forecast. According to Shell, "A scarcity of oil supplies—including unconventional sources and natural gas liquids—is very unlikely before 2025. This could be extended to 2040 by adopting known measures to increase vehicle efficiency and focusing oil demand on this sector." As we will see, Shell's hedges—"unconventional sources and natural gas liquids" as well as "known measures" to increase vehicle efficiency—will largely decide how the peaking of global oil production will affect the climate and our way of life.

"Conventional" oil means the liquid crude petroleum that is extracted from the ground using the traditional method, an oil well. Experts do not agree on how much "ultimately recoverable" oil remains in the ground, in part because they use different definitions for conventional oil and in part because they disagree about how much technology advances will enable more oil to be found and extracted. Also, the peak-oil "pessimists" simply don't believe the claims by some Middle East governments as to how much conventional-oil reserves they have.

Resolving that dispute is beyond the scope of this book, but a few points are critical to understand. The vast majority of the world's conventional-oil reserves are in unstable regions, such as the Middle East, guaranteeing extreme oil-price volatility for decades to come. The rapid growth in demand for oil by developing countries, especially China and India, coupled with the refusal by the United States to adopt strong policies to restrain or reverse our own growing demand, ensure that conventional oil will almost certainly peak and then decline sometime in the next quarter-century. The world is in fact running out of conventional oil. What about unconventional oil?

PEAK OIL AND GLOBAL WARMING

Unfortunately, *most forms of unconventional oil will make global warming worse*—and some of them will make Hell and High Water all but inevitable. Ironically, global warming is making it easier to explore and drill for oil in the Arctic because the sea ice is vanishing at an ever-increasing rate. The amount of undiscovered oil in the Arctic has been estimated at 200 to 400 billion barrels—enough to supply the world for seven to fourteen years at current usage. Let's look at some of the major unconventional sources.

First, we have a number of viscous oils called bitumen, heavy oil, and tar sands (or oil sands). There is more recoverable heavy oil in Venezuela than there is conventional oil in Saudi Arabia, and Canada has even more recoverable oil in its tar sands. Tar sands are pretty much the heavy gunk they sound like, and making liquid fuels from them requires huge amounts of energy for steam injection and refining. Canada is currently producing about a million barrels of oil a day from the tar sands, and that is projected to triple over the next two decades.

The tar sands are doubly dirty. On the one hand, the energy-intensive conversion of the tar sands directly generates two to four times the amount of greenhouse gases per barrel of final product as the production of conventional oil. On the other hand, Canada's increasing use of natural gas to exploit the tar sands is one reason that its exports of natural gas to the United States are projected to shrink in the coming years. So instead of selling clean-burning natural gas to this country, which we could use to stop the growth of carbon-intensive coal generation, Canada will provide us with a more carbon-intensive oil product to burn in our cars. That's lose-lose.

From a climate perspective, fully exploiting the tar-sands resource would make Canada's climate policy as immoral as ours.

Second, even more oil can probably be recovered from shale, a claylike rock, than from the tar sands. Most of the world's shale is found in the United States, and most of our shale, a trillion tons, is found in Colorado and Utah. After the oil shocks of the 1970s, billions were spent exploring the possibility of shale oil, but those efforts were abandoned in the 1980s when oil prices collapsed. Shale does not contain much energy—per pound, it has one-tenth the energy of crude oil and one-fourth that of recycled phone books. Converting shale to oil requires a huge amount of energy—possibly as much as 1,200 megawatts of generating capacity to produce only 100,000 barrels per day. If those were fossil-based megawatts, we would be spewing millions of tons of greenhouse gases into the air every year just to create a fuel that itself would spew more greenhouse gases into the air when burned in a car. But then it would be equally crazy to use renewable energy to make shale, when we critically need that zero-carbon power to displace coal electricity.

We simply must leave the shale in the ground.

Third, the recovery of conventional oil from a well can be enhanced by injecting carbon dioxide (CO_2) into the reservoir. Estimates for potential recovery are 300 to 600 billion barrels. A 2005 study, "Peaking of World Oil Production," led by Science Applications International Corporation (SAIC) explained:

> CO_2 flooding can increase oil recovery by 7–15 percent of original oil in place. Because EOR (enhanced oil recovery) is relatively expensive, it has not been widely deployed in the past. However, as a way of dealing with peak conventional oil production and higher oil prices, it has significant potential.

The SAIC study might also have noted that in a world where carbon capture and storage from coal generation becomes commonplace—which might occur as soon as two decades from now—we may be awash in carbon dioxide that could be diverted to EOR. What a double tragedy it would be if that carbon dioxide were not put into deep underground aquifers (permanently reducing the amount of heat-trapping gas in the atmosphere) but instead was used to extract more fossil fuels from the ground (which would ultimately release carbon dioxide into the atmosphere when burned in internal combustion engines).

Fourth, coal and natural gas can be converted to diesel fuel using the Fischer-Tropsch process. During World War II, coal gasification and liquefaction produced more than half of the liquid fuel used by the German military. America has so much coal, it could replace all imported oil with liquid fuel from coal—and keep generating electricity from coal—for more than 100 years. China has nearly as much coal as we do. The Chinese are launching a huge coal-liquefaction effort and plan to generate 300,000 barrels of oil a day from coal by 2020.

The process is incredibly expensive. You need to spend $5 billion or more just to build a plant capable of producing only 80,000 barrels of oil a day (the United States currently consumes more than 21 million barrels a day). You need about 5 gallons of water for every gallon of diesel fuel that's produced—not a particularly good long-term strategy in a world facing mega-droughts and chronic water shortages. Worse, the total carbon dioxide emissions from coal-to-diesel are about double that of conventional diesel. You can capture the carbon dioxide from the process and store it underground permanently. But that will make an expensive process even more expensive, so it seems unlikely for the foreseeable future, certainly not until carbon dioxide is regulated and has a high price, and we have a number of certified underground geologic repositories.

More important, even if you capture the CO_2 from the Fischer-Tropsch process, you are still left with diesel fuel, a carbon-intensive liquid that will release carbon dioxide into the atmosphere once it is burned in an internal combustion engine. A great many people I have spoken to are confused about this point; they think that capturing and storing the CO_2 while turning coal to diesel is as good an idea as capturing the CO_2 from the integrated gasification combined-cycle process for turning coal into electricity. No. The former process still leaves you with a carbon-intensive fuel, whereas the latter process leaves you with zero-carbon electricity. Worse, some people propose taking the captured CO_2 and using it for enhanced oil recovery, which, as we've seen, is the equivalent of not capturing the carbon dioxide at all.

Coal-to-diesel is a bad idea for the planet. If the United States or China pursues it aggressively, catastrophic climate change will be all but unavoidable. Turning natural gas into diesel is not as bad an idea, at least from the perspective of direct emissions, because natural gas is a low-carbon fuel. But it represents a tremendous misuse of natural gas, which could otherwise be used to displace coal plants and sharply reduce future greenhouse gas emissions.

A 2006 study by the University of California at Berkeley found that meeting the future demand shortfall from conventional oil with unconventional oil, especially coal-to-diesel, could increase annual carbon emissions by 2 billion metric tons (7.3 gigatons of carbon dioxide) for several decades. That would certainly be fatal to any effort to avoid 20 to 80 feet of sea-level rise.

We are simply running out of time, and we no longer have the luxury of grossly misallocating capital and fuels. That's why significantly exploiting unconventional sources of liquid fossil fuel such as coal, tar sands, and shale is the road to ruin. And that's why the Bush administration efforts to push hydrogen-fuel-cell cars make so little sense.

THE HYPE ABOUT HYDROGEN

> *Forget hydrogen, forget hydrogen, forget hydrogen.*
> —Former CIA director James Woolsey, 2006

The promise of hydrogen cars as a simple techno-fix, a deus ex machina to solve our environmental problems painlessly, and without regulations, is a cornerstone of the Bush administration's climate policy. In his January 2003 State of the Union address, the president pledged "$1.2 billion in research funding so that America can lead the world in developing clean, hydrogen-powered automobiles." He then said that "the first car driven by a child born today could be powered by hydrogen, and pollution-free."

The president didn't tell the public that more than 98 percent of the hydrogen made in this country today must be extracted from fossil fuel hydrocarbons—natural gas, oil, and coal—and that process releases huge amounts of carbon dioxide. "It is highly likely that fossil fuels will be the principal sources of hydrogen for several decades," concluded a prestigious National Academy of Sciences panel in 2004. In fact, hydrogen as a transport fuel might even *in-*

crease greenhouse gas emissions rather than reduce them. That was the conclusion of a January 2004 study by the European Commission and European oil companies and car companies.

The only way hydrogen cars could be "pollution-free" is for the hydrogen to be made from pollution-free sources of energy, like wind power. But the administration and Congress won't even pass a law requiring that 10 percent of U.S. electricity be renewable by 2020—so what are the chances that children born in 2003 will be driving a car in 2020 with pollution-free hydrogen?

Making hydrogen for use in cars is a terrible use of pollution-free power. Instead, we should build renewable-power plants to avoid the need to build new coal plants and save *four times as much carbon dioxide at less than one-tenth the cost* of using that same renewable power to make hydrogen to run a car. Study after study has shown that it makes no sense to squander renewable power to make hydrogen for cars until the electric grid is itself virtually greenhouse-gas-free—and that is at least four decades away. That's 40 years from now, even if we are able to reverse our current energy policy the day after Bush leaves office.

And this analysis assumes that hydrogen cars will actually become practical for consumers any time soon. But that is highly unlikely. They simply require too many scientific breakthroughs. For starters, a pollution-free hydrogen car requires a fuel cell for efficiently converting hydrogen into useful energy without generating pollution. Fuel cells are small, modular electrochemical devices, similar to batteries, except that they can be continuously fueled. They take in hydrogen and oxygen and put out only water plus heat and electricity, which runs an electric motor.

Unfortunately, fuel cells for cars currently cost about $2,000 per kW, which is about fifty times greater than the cost of an internal combustion engine. A major breakthrough will be required to make fuel cells affordable and practical.

Yet another major breakthrough is needed to solve the storage problem. Hydrogen is the most diffuse gas there is. No known material can store enough of it in a practical way to give people the kind of driving range they want. A March 2004 study by the American Physical Society concluded that "a new material must be discovered" to solve the storage problem.

Another problem: Currently hydrogen from pollution-free renewable sources would cost the equivalent of $6 to $10 a gallon of gasoline. So we'll need another major breakthrough that will drop the cost by a factor of three.

Hydrogen cars need three major breakthroughs—in fuel cells, storage, and renewable hydrogen—within the next decade or so, in a world where game-changing energy-technology breakthroughs hardly ever happen. And if those three happened, we would still need someone to spend more than $500 billion to build the fueling infrastructure needed to make hydrogen available throughout the country. An analysis in the May 2004 issue of *Scientific American* stated, "Fuel-cell cars, in contrast [to hybrids], are expected on about the same schedule as NASA's manned trip to Mars and have about the same level of likelihood."

When Bill Reinert, the U.S. manager of Toyota's advanced technologies group, was asked in 2005 when fuel-cell cars would replace gasoline cars, he answered, "If I told you '*never*,' would you be upset?" A 2004 MIT study concluded that hydrogen-fuel-cell cars would be unlikely to achieve significant market success until the year 2060, far too late to help.

And yet in spite of all this, the Bush administration keeps pumping money into the budget for hydrogen. In its 5-year budget outlook released in 2006, the hydrogen-technology budget rose to a stunning $323 million in fiscal year 2011 (out of $1.13 billion for all energy efficiency and renewable energy) from a requested $196 million in 2007 (out of $1.18 billion). The tragedy of this is a 20 percent drop in funding for technologies that actually hold some promise of helping to reduce greenhouse gas emissions in the first half of this century.

In April 2005, Energy Secretary Samuel Bodman announced that he was disbanding the department's primary independent advisory board on scientific and technical matters, a board that has existed in some form since 1978. Bodman is uninterested in outside scientific advice. A department spokesman claimed Bodman was doing this because he is a chemical engineer by training and "the secretary has an understanding of science and scientific processes." But Bodman's 5-year budget plan grossly misdirects more than a billion dollars of the department's research-and-development funds, suggesting that he doesn't understand at all.

THE WIN-WIN OIL POLICY

> *My message is that the balance of realism has passed from those who argue on behalf of oil and a laissez-faire energy policy that relies on market evolution, to those who recognize that in the absence of a major reorientation in the way we get our energy, life in America is going to be much more difficult in the coming decades.*
>
> —Senator Richard Lugar, 2006

If neither hydrogen cars nor the peak and subsequent drop in global oil production are going to save us from endlessly rising greenhouse gas emissions, what will?

I have described a variety of aggressive low-carbon strategies or "wedges" we need to achieve over the next five decades. Each wedge ultimately avoids 1 billion metric tons of carbon emissions a year.

The last chapter looked at the five wedges needed to reduce

émissions from electricity, buildings, and heavy industry. The two wedges needed in the transportation sector are:

- Every car and SUV achieves an average fuel economy of 60 miles per gallon.
- Every car can run on electricity for short distances before reverting to biofuels.

How do we ensure that the *average* car on the road in 2060 gets 60 mpg, when the current average is about one-third that? Some push for high gasoline taxes. European countries such as the United Kingdom and Germany have taxes of more than $2 per gallon, which is five times more than the U.S. tax. Yet the average fuel economy of European Union vehicles is nowhere near 60 mpg. Oil and gasoline prices will probably trend higher over the next two decades by themselves as demand continues to grow in the face of supply constraints, and as terrorism and instability cause price spikes and oil-market jitters. When this country gets serious about global warming, we will put in place a carbon-trading system that will increase the price of gasoline somewhat, though far less than European gas taxes do today. I don't think higher gas taxes are the best way to get 60-mpg cars.

Another, more obvious strategy is tougher fuel-efficiency standards. After all, corporate average fuel economy (CAFE) standards, enacted in 1975, were used successfully in this country to double the fuel efficiency of our cars while making them safer, mandating that new cars have a fuel efficiency of 27.5 miles per gallon. In a 2002 report to President Bush, the National Academy of Sciences concluded that automobile fuel economy could be increased by up to 42 percent for large SUVs with technologies that would pay for themselves in fuel savings. That study did not even consider the greater use of diesels and hybrids. The report was ignored.

Studies by the national laboratories, by MIT, and by the Pew

Center on Global Climate Change have concluded that even greater savings could be cost-effective while maintaining or improving passenger safety. In a comprehensive 2005 study of fuel economy and traffic fatalities in industrialized nations, Robert Noland of the Centre for Transport Studies at Imperial College in London found that "average fleet fuel economy has no effect on traffic fatalities." The conclusion: "Policies aimed at improving fuel economy," whether to reduce dependence on imported oil or to reduce carbon dioxide emissions, "will most likely not have adverse safety consequences." Indeed, greater use of hybrid technology should *increase* vehicle safety. *Automotive Engineering International*, which named the Toyota Prius hybrid "Best Engineered Vehicle 2004," explained that the Prius has a variety of safety features, including an electronic brake-by-wire system and a skid-control computer that coordinates with the hybrid system control computers. Hybrid electronics hold the promise of far more controllability, quicker response, and greater safety.

Even with their much higher gasoline prices, the Europeans have still insisted on a voluntary agreement with automakers to further reduce carbon dioxide emitted per mile by about 25 percent from 1996 to 2008 for the average light-duty vehicle, which equates to a vehicle fuel efficiency of about 44 mpg. Japan has a mandatory target with similar goals. Even China has a far tougher standard than we do, plus a 20 percent tax on the most inefficient cars. The car of the future is definitely fuel-efficient.

Our own federal law is a large obstacle—it still requires that the average new car get 27.5 miles per gallon (the same level we had in 1985). The average SUV must get a mere 20.7 miles per gallon (up a tad from 19.5 mpg in 1985). Worse, the National Highway Traffic Safety Administration uses data from unrealistic tests of vehicle efficiency to judge how well car companies have met the CAFE standard. The result is that in 2005, *Consumer Reports* found that the fleet of 2003-model passenger cars they tested averaged only a pitiful 22.7 mpg, far below the 27.5 the law requires (and even farther below the 29.7 mpg that National Highway Transportation Safety Administration had somehow calculated for those models). The light trucks they tested measured a meager 16 mpg, far below the law's 20.7.

We could design the standards more flexibly, and many groups, including the bipartisan National Commission on Energy Policy, have suggested improvements to CAFE. In 2005 the Center for American Progress proposed that the government offer to help U.S. car companies with their legacy health-care costs in return for a commitment to steady improvements in vehicle fuel efficiency. The climate challenge is so enormous that we will certainly need creative deals and bargains like that if we are to have any chance of avoiding catastrophe.

Another worthwhile strategy would be vehicle standards that reduce carbon dioxide emissions from the tailpipe, which in the short run would increase vehicle efficiency but in the long run would include low-carbon alternative fuels. California has put forward just such a carbon dioxide standard, and ten other states have followed. Tragically, those standards have been strongly opposed by both the Bush administration and the auto companies.

In 2006, Bush did slightly increase the fuel-economy standards for SUVs, and included huge gas-guzzling SUVs that exceed 8,500 pounds, such as GM's Hummer H2, which had previously been exempt from such regulations. But the change was minor and left open a huge loophole that exempts large pickup trucks. In the year 2025, the new standards will save the nation about *two weeks' worth of oil.* Hardly a treatment for a serious addiction. Also, the change appears to have been introduced not to have cleaner energy but to allow the administration to better argue in court that its new federal standards preempt California's much stronger proposed standards.

All that said, requiring improved vehicle efficiency, by itself, cannot achieve the greenhouse gas reductions we will need—

because if the world's population and economies continue their rate of growth, the number of cars on the road will triple by mid-century. So we will also need one or more zero-carbon alternative transportation fuels. Those alternative fuels will have to be electricity and biofuels.

THE CAR AND FUEL OF THE FUTURE

With a straightforward improvement to current hybrids, they can be plugged in to the electric grid and run in an all-electric mode for a limited range between recharging. If the initial battery charge runs low, these plug-in hybrids can run solely on gasoline.

We Americans use our cars mainly for relatively short trips, such as commuting—half of American cars travel less than 30 miles a day—followed by extended periods when the vehicle could be re-charged. So an all-electric range of 20 to 30 miles would allow these plug-in hybrid vehicles to replace a substantial portion of gasoline consumption and tailpipe emissions. If the electricity came from CO_2-free sources, these vehicles would dramatically reduce net greenhouse gas emissions.

A conventional car costs about 12 cents a mile to operate, for gasoline costing $2.50 a gallon. In contrast, a plug-in hybrid could run on electrons at 3 cents a mile, using electricity that costs about 8 cents a kilowatt-hour, the current average residential rate. Battery improvement—especially the next generation of lithium-ion batteries that will be available by 2010—will lead to increased functionality for plug-in hybrids. The larger battery of a plug-in hybrid, coupled with a higher-powered electric motor, allows significant downsizing of the gasoline engine and other related mechanical systems. Engineers at the University of California at Davis have built several plug-in hybrid prototypes that can travel 60 miles on electricity alone, with engines that are less than half the size of standard engines.

Plug-in hybrids avoid many of the barriers that have plagued alternative-fuel vehicles and that make hydrogen-fuel-cell cars so impractical. Plug-in hybrids do not have a limited range. They do not have a high fueling cost compared with gasoline. In fact, the per-mile fueling cost of running on electricity is about one-third the per-mile cost of running on gasoline. The key infrastructure dilemma—who will build the new hydrogen-fueling infrastructure until the cars are a success, but who will buy the cars if there aren't thousands of fueling stations already built—is minimized because electricity is widely available and charging is straightforward.

The plug-in hybrid will have a higher first cost, but this will be paid back by the lower fuel bill. One 2006 study found that with gasoline at $3 a gallon—probably the low end of the price range by the time we could begin a broad transition to plug-ins in a decade—the extra cost of the vehicle will be returned in five years, even if electricity prices rise 25 percent from current levels.

The remarkably low fueling cost of the best current hybrids (like the Toyota Prius) and future plug-in hybrids are a major reason why I don't worry as much about peak oil as some do. James Kunstler, for instance, argues in his 2005 book, *The Long Emergency,* that after oil production peaks, suburbia "will become untenable" and "we will have to say farewell to easy motoring." But suppose Kunstler is right. Suppose oil hits $160 a barrel and gasoline goes to $5 a gallon in, say, 2015. That price would still be lower than many Europeans pay today. You could just go out and buy the best hybrid and cut your fuel bill in half, back to current levels. Hardly the end of suburbia. And suppose oil hit $280 a barrel and gasoline rose to $8 dollars a gallon in 2025. You would replace your hybrid with a plug-in hybrid, and those trips under 30 miles that have made suburbia what it is today would actually *cut* your fuel bill by a factor of more than ten—even if all the electricity were from zero-carbon sources like wind power—*to far below what you are paying today.*

I expect commercial plug-in hybrids to be available within a few years. And as battery technology improves and gasoline prices rise in the coming decade, plug-ins will become increasingly popular. Growing concern over global warming will only serve to accelerate the transition.

THE CAR OF THE FUTURE IS CLIMATE-FRIENDLY

Environmentally, plug-in hybrids have an enormous advantage over hydrogen-fuel-cell vehicles in utilizing zero-carbon electricity because of the inherent inefficiency of generating hydrogen from electricity, transporting hydrogen, storing it aboard the vehicle, and then running it through the fuel cell. The overall efficiency of a hydrogen-fuel-cell vehicle's ability to use renewable electricity is a meager 20 to 25 percent. The efficiency of charging an onboard battery and then discharging it to run an electric motor in a plug-in hybrid, however, is 75 to 80 percent.

Replacing half of the U.S. ground-transport fuels (gasoline and diesel) with hydrogen from wind power by 2050, for example, might require 1,400 gigawatts of advanced wind turbines or more. However, replacing those fuels with electricity (for plug-in hybrids) might require less than 400 GW. That 1,000-GW difference is an insurmountable obstacle for hydrogen fuel especially because the United States will need hundreds of gigawatts of wind and other zero-carbon power sources just to sharply reduce greenhouse gas emissions in the electricity sector, as we have seen.

Another advantage of plug-ins is that they hold the potential to make intermittent renewable power, like wind, more cost-effective. Wind delivers power only when the wind is blowing, and this is not as valuable to electric utilities as base-load power plants that provide power available all the time. But most cars stay parked for more than twenty hours a day. We can imagine that an electric utility might lease a plug-in hybrid to a consumer or business willing to leave the vehicle connected when it was not on the road and to permit the utility to control when the vehicle's battery was charged and discharged depending on its generation or voltage-regulation needs. Such an arrangement would help utilities with load balancing. It would also allow utilities to do most of the charging when the wind was blowing, eliminating the need for costly electricity storage that high levels of wind power might otherwise need. One reason the municipal utility Austin Energy has helped launch a national campaign for the plug-in hybrid is that they have so much West Texas wind power available at night.

CELLULOSIC ETHANOL

Biomass can be used to make a zero-carbon transportation fuel, such as ethanol, which is now used as a gasoline blend. Today, the major U.S. biofuel is ethanol made from corn, which yields only about 25 percent more energy than was consumed to grow the corn and make the ethanol. A considerable amount of R&D is being spent on producing ethanol that can be made from far less energy-intensive sources. Called cellulosic ethanol, it can be made from agricultural and forest waste and also from dedicated energy crops, such as switchgrass or fast-growing poplar trees, which can be grown and harvested with minimal energy consumption, so that overall net emissions are near zero.

Ethanol's advantage over alternative fuels like hydrogen gas is that it is a liquid fuel and thus much more compatible with our existing fueling system. Existing oil pipelines, however, are not compatible with ethanol, so significant infrastructure spending would still be required before ethanol could become the major transportation fuel. And ethanol production will require technological advances before it can match the price of (untaxed) gasoline on an equivalent energy basis. Carnegie Mellon University researchers note that cellulosic ethanol costs the equivalent of "$2.70 per gallon in order to get as much energy as in a gallon of gasoline."

Thus, if oil prices in, say, 2020 are consistently higher than $70 a barrel, cellulosic ethanol could be a competitive alternative fuel. This is particularly true because by that time we will inevitably have a price for carbon, further improving the cost of cellulosic ethanol relative to gasoline.

Probably the biggest barrier to biofuels, and to biomass energy in general, is that biomass is not very efficient at converting and storing solar energy, so large land areas would be needed to plant enough crops to provide a significant share of transportation energy. One 2001 analysis by ethanol advocates concluded that to provide enough ethanol to replace the gasoline used in the light-duty fleet alone, "it would be necessary to process the biomass growing on 300 million to 500 million acres, which is in the neighborhood of one-fourth of the 1.8 billion acre land area of the lower 48 states" and is roughly equal to the total amount of U.S. cropland in production today. That amount of displaced gasoline represents about 60 percent of all U.S. transportation-related carbon dioxide emissions today but under 40 percent of what is projected for 2025 under a business-as-usual scenario. Given the vast acreage needed, using so much land for fuel would obviously have dramatic effects—environmental, political, and economic.

If ethanol is to represent a major transportation fuel in the coming decades, then U.S. vehicles will need to become far more fuel-efficient. A fleet of 60-mpg cars would substantially reduce the biomass acreage requirements. And putting cellulosic ethanol blends into plug-in hybrids would further reduce acreage requirements, especially since there are plausible strategies for cogeneration of biofuels and biomass electricity.

In the long term, biomass-to-energy production could be exceedingly efficient with "biorefineries" that produce multiple products. Dartmouth engineering professor Lee Lynd described one such future biorefinery where cellulosic ethanol undergoes a chemical pretreatment, then fermentation converts the carbohydrate content into ethanol, as CO_2 bubbles off. The residue is mostly lignin (a polymer found in the cell walls of plants). Water is removed, and the biomass residue is then gasified to generate electricity or to produce a stream of hydrogen and CO_2. The overall efficiency of converting the energy content of the original biomass into useful fuel and electricity would be an impressive 70 percent, even after accounting for the energy needed to grow and harvest the biomass. The CO_2 can be sequestered. Also, this process could be used to generate biodiesel. This is a futuristic scenario, one that is the subject of intense research and that could make ethanol directly competitive with gasoline, and biomass electricity competitive with other zero-carbon alternatives, especially when there is a price for reducing CO_2 emissions.

ENERGY SECURITY AS A SIDE BENEFIT

Because of the abundance of unconventional oil and low-cost alternative fuels, peak oil is not the major energy problem that threatens the American way of life. Yes, if we don't aggressively pursue fuel efficiency and low-carbon alternative fuels *now*, the nation certainly faces oil price shocks and steadily increasing prices over the next quarter-century. But if we fail to pursue those crucial strategies, then Planetary Purgatory and 20-foot sea-level rise becomes all but inevitable, and we face the multidecade struggle to avoid the worst of Hell and High Water. Even if conventional oil peaks within two decades, the growing use of dirty, unconventional oil, along with rapidly rising natural-gas and coal consumption, will generate far too much carbon dioxide.

Global warming is the energy problem that threatens the American way of life. Over the next few decades, we need to triple the efficiency of our cars and SUVs, and have them also be flexible-fuel plug-in hybrids that run mostly on zero-carbon electricity and cellulosic ethanol. Whether your primary concern is peak oil and our energy insecurity or global warming, it is important to recognize that sharply reducing our reliance on oil will not happen with the Bush administration strategy. Their strategy is *rhetoric* about our oil addiction plus the *reshuffling* of some of our federal R&D dollars while at the same time blocking national efforts to boost the use of renewable power and opposing state efforts to boost vehicle fuel efficiency.

Triple-efficiency vehicles will be the norm by 2050 only with much higher mileage standards of the kind that most other countries, including China, are embracing (or with tailpipe-emissions standards for carbon dioxide, as California and ten other states propose). If we fail to embrace such standards nationally, the rest of the world will lead in advanced automotive technology, and GM and Ford will continue to bleed market share and jobs. The standard should be written in such a way as to encourage companies to meet them with hybrid technology, because that will help make cars safer and jump-start the shift to plug-in hybrids.

A successful transition to alternative fuels also requires government standards. Indeed, the only reason Brazil has been so successful in replacing gasoline with ethanol is that the government required minimum levels of ethanol blends and then required all gasoline stations to have at least one ethanol pump. We need such sensible policies in the United States. Here are two from the National Commission on Energy Policy:

- Develop the first six pioneer cellulose-to-energy plants between 2008 and 2012 using production or investment incentives.
- Modify agricultural subsidies to include energy crops without increasing total farm subsidies or decreasing farm income.

We should sharply increase federal investments in biofuels and advanced batteries while cutting the hydrogen program by more than half. We should adopt a renewable-fuels standard whereby 25 percent of our gasoline would be replaced with cellulosic ethanol by 2025. We should also launch a major effort to have at least 10 percent of our new cars be plug-in hybrids by 2025.

These strategies would not only sharply reduce greenhouse gas emissions from cars but would do so without raising the nation's fuel bill for transportation. As huge side benefits, we could achieve genuine energy security, sharply lower our trade deficit, revitalize our domestic auto industry, create countless jobs, and increase our national security, because we would no longer be beholden to un-democratic governments in the Middle East or have our economy repeatedly subject to price shocks from political instability or ter-rorist attacks.

CHAPTER NINE

THE U.S.–CHINA SUICIDE PACT ON CLIMATE

*The "international fairness" issue is the emotional
home run. Given the chance, Americans will demand
that all nations be part of any international global
warming treaty. Nations such as China, Mexico and
India would have to sign such an agreement for the
majority of Americans to support it.*

—Frank Luntz, 2002

*We don't need an international treaty with rules and
regulations that will handcuff the American economy
or our ability to make our environment cleaner, safer
and healthier.*

—Frank Luntz, 2002

What country's insatiable thirst for oil imports is most responsible for the tightening world market since the mid-1990s?
Hint: It's not China. From 1995 to 2004, China's annual imports
grew by 2.8 million barrels a day. Ours grew 3.9 million. China now
sucks up about 6 percent of all global oil exports. We demand 25
percent, even though China has a billion more consumers.

In what year will China's total contribution to climate change
from burning fossil fuels surpass ours? Hint: Climate change is

driven by rising atmospheric concentrations of greenhouse gases, and those concentrations have been driven by cumulative emissions since the dawn of the industrial revolution. While China's *annual* CO_2 emissions may well exceed ours by 2025, its *cumulative* emissions might not surpass ours until after 2050.

Not only are we the richest nation in the world, but for many decades to come we will be the one most responsible for global warming. No wonder the Chinese and Indians and others in the developing world expect us to take action first, just as we did to save the ozone layer. No wonder the rest of the industrialized world embraced the Kyoto restrictions on greenhouse gas emissions, even knowing the emissions from developing countries such as China and India were not restricted.

One can only marvel at a strategist like Frank Luntz for his ability to appeal to Americans who "will demand that all nations be part of any international global warming treaty," while, in the same breath, reaching out to Americans who oppose "an international treaty with rules and regulations that will handcuff the American economy." Such a rhetorical flimflam strategy by the global-warming Denyers and Delayers is politically very savvy, but it is the sure road to Hell and High Water.

That said, China's emissions *are* growing at an alarming rate. In 2000 the government walked away from the California-style energy-efficiency effort it had embraced since 1980. For the past few years, it has been building one major dirty-coal plant *almost every week.* The climate problem cannot be solved if China and other rapidly developing countries do not take steps to restrain their emissions growth. But if the United States maintains its position that we will not take strong action until China does, neither country is likely to act in time. This chapter explores how the United States and China might avoid destroying the climate and, with it, our way of life.

A BRIEF HISTORY OF TIMETABLES

> *Perhaps the most extraordinary aspect of the Montreal Protocol [on Substances that Deplete the Ozone Layer] was that it imposed substantial short-term economic costs in order to protect human health and the environment against speculative future dangers—dangers that rested on scientific theories rather than on proven facts. Unlike environmental agreements of the past, this was not a response to harmful developments or events, but rather preventive action on a global scale.*
>
> —Richard Benedick, former ambassador, 2005

The ozone layer shields life on Earth from the sun's harmful ultraviolet rays. In 1974, climate scientists warned us that chlorofluorocarbons (CFCs) were destroying Earth's ozone layer, threatening to bring about a sharp increase in skin cancer. Within only 5 years, the United States voluntarily banned their use in spray cans, and CFC production began to decline. But other uses for CFCs, as refrigerants and solvents, began driving the demand up again by the early 1980s.

In 1985, scientists discovered a hole in the ozone shield over Antarctica. As the National Academy of Sciences wrote, this was "the first unmistakable sign of human-induced change in the global environment. . . . Most scientists greeted the news with disbelief. Existing theory simply had not predicted it."

Chlorine concentrations had been increasing over Antarctica for decades, up from the natural level of 0.6 parts per billion. Yet as Richard Benedick, President Ronald Reagan's chief negotiator at the Montreal conference, explained in a 2005 Senate hearing, "no effect on the ozone layer was evident until the concentration exceeded two parts per billion, which apparently triggered the totally unexpected collapse." His ominous lesson for today: "Chlorine concentrations had *tripled* with no impact whatsoever on ozone until they crossed an unanticipated threshold." As we have seen repeatedly, Earth's climate system has many such thresholds.

The stunning revelation of an ozone hole drove the world to negotiate the Montreal Protocol. The 1987 agreement called for a 50 percent cut in CFC production by 1999. Significantly, the protocol's targets and timetables *slowed the rate of growth of concentrations only slightly* and would still have led to millions of extra skin cancer cases by midcentury. Further, the protocol allowed developing countries to delay implementing the control measures for about ten years and required rich countries to give them access to alternative chemicals and technologies together with financial aid.

Nevertheless, President Reagan endorsed the protocol, and the Senate ratified it. By the end of 1988, twenty-nine countries and the European Economic Community—but not China or India—had ratified it. The treaty came into effect the next year, but it took many more years of negotiations, continuous strengthening of the scientific consensus, and significant concessions to developing countries in both technology transfer and financial assistance, before amendments to the treaty were strong enough and had enough support from rich and poor countries alike to ensure that CFC *concentrations* in the air would be reduced.

The analogy of the ozone layer and the Montreal Protocol to global warming and the Kyoto Protocol is far from perfect—greenhouse gases are more integral to modern life than CFCs ever were. American politics have changed in two decades, and the terms of the Montreal Protocol would no doubt be viewed today as wholly inadequate and politically unacceptable, especially without ratification by China and India. Yet this small first step by the rich nations jump-started a multiyear process that saved the ozone layer and prevented millions of cases of skin cancer. It also set an example of how the world could come together to tackle the climate problem.

For many decades, scientists have been warning us about the dangers of greenhouse gases. By the late 1970s, the National Academy of Sciences, the nation's most prestigious scientific body, had warned that uncontrolled greenhouse gas emissions might raise global temperatures 10°F and cause sea levels to rise catastrophically. The discovery of the ozone hole in 1985—an unexpected climate impact from an unanticipated emissions threshold—made us more aware of how we have affected the climate and helped push the nations of the world into an international effort to control greenhouse gas emissions.

In 1992, President Bush's father signed the United Nations Framework Convention on Climate Change (UNFCCC), also called the Rio climate treaty, and that year the Senate ratified it unanimously. The convention's goal was to set up an international process to stabilize "greenhouse gas concentrations in the atmosphere at a level that would prevent dangerous anthropogenic [human-made] interference with the climate system." The UNFCCC did not establish what that level was but did establish a nonbinding target that called for developed countries to return their emissions of greenhouse gases to 1990 levels. Perhaps most significant, the signatories to the treaty recognized "that the largest share of historical and current global emissions of greenhouse gases has originated in developed countries, that per capita emissions in developing countries are still relatively low and that the share of global emissions originating in developing countries will grow to meet their social and development needs." The Rio treaty recognized the "common but differentiated responsibilities and respective capabilities" of each nation and established a core principle: "Accordingly, *the developed country Parties should take the lead in combating climate change* and the adverse effects thereof" (emphasis added).

Unfortunately, supporters of action on climate change, including those in the Clinton administration, never fully explained to the American people how and why the rich countries had promised to take the lead in combating climate change. As a result, the U.S. Senate passed a resolution in 1997, offered by Senators Robert Byrd (Democrat) and Chuck Hagel (Republican), with a vote of 95–0, stating a "sense of the Senate" that the United States should not sign any protocol to the UNFCCC that would "mandate new commitments to limit or reduce greenhouse gas emissions for the [industrialized countries], unless the protocol or other agreement also mandates new specific scheduled commitments to limit or reduce greenhouse gas emissions for Developing Country Parties within the same compliance period."

Probably the Clinton administration's biggest political mistake on the climate issue was making no serious effort to stop that 95–0 outcome. This meant that the 1997 Kyoto Protocol, which set targets and timetables only for the emissions of rich countries, was dead before it got to the U.S. Senate—even though it was similar in most important respects to the Montreal Protocol, which had passed the Senate a decade earlier and had saved the ozone layer and the lives of countless Americans.

When you talk to people from China, India, or other developing countries, they don't understand our politics at all. They don't understand how the country that became the richest by spewing greenhouse gases that are now destroying everybody's climate would refuse to use some of that wealth to prevent catastrophic warming. They find it absurd that American politicians argue for delay by saying we must wait for the poorest countries to make commitments at the same time.

In the 1997 Byrd-Hagel amendment that helped kill the protocol, the senators stated their objection: "whereas greenhouse gas emissions of Developing Country Parties are rapidly increasing and are expected to surpass emissions of the United States and other

OECD countries as early as 2015." That language sounds so reasonable. As Luntz wrote, *"The 'international fairness' issue is the emotional home run."*

But remember that the key metric is not *annual* emissions but *cumulative* emissions. Cumulative emissions are what drive up carbon dioxide concentrations, and concentrations are what determine how much the planet warms. Developed countries had four times the cumulative emissions of developing countries from 1850 to 1995. The rich countries' total emissions from fossil fuel consumption would exceed that of the poor countries through midcentury. Even in the year 2000, the average American emitted nine times the carbon dioxide of a typical Chinese and twenty times that of a typical Indian. And, of course, the rich countries were (and still are) far, far, richer, especially on a per capita basis. That's why few developing countries are likely to agree to serious restrictions on their greenhouse gas emissions until and unless the developed countries go first, which is what we agreed to under the Rio treaty. And that's why virtually every developed country (other than the United States) agreed to the terms of the Kyoto Protocol.

THE CHINA SYNDROME

China's energy history can be divided into several phases, as we learn from Dr. Mark Levine, cofounder of the Beijing Energy Efficiency Center. The first phase (1949–1980) was a "Soviet-style" energy policy characterized by subsidized energy prices, no concern for the environment, and energy use that rose faster than economic growth (GDP).

The second phase (1981–1999) was "California on steroids," when the country embraced an aggressive push on energy management and energy efficiency, surpassing the efficiency efforts California has achieved since the mid-1970s. This came about as a result of Deng Xiaoping heeding the advice of leading academic experts who suggested a new approach to energy. Chinese strategies included

- factory energy-consumption quotas and energy-conservation monitoring
- efficient technology promotion and closing of inefficient facilities
- controls on oil use
- low interest rates for efficiency-project loans
- reduced taxes on efficient-product purchases
- incentives to develop new efficient products
- monetary awards to efficient enterprises
- strategic technology development and demonstration
- national, local, and industry-specific technical efficiency service and training centers

During the mid-1990s, China also began dramatic energy-price reforms, which led to higher prices for coal, oil, and electricity. China's policies kept energy growth to a modest level during a time of explosive economic growth. For instance, from 1990 to 2000, its economy more than doubled, but carbon dioxide emissions rose by only one-fourth. Remarkably, *during the 1990s, the United States actually increased its annual emissions of carbon dioxide more than China did.*

Unfortunately, toward the end of the last decade, China scaled back or eliminated many of its efficiency efforts, leading to the third phase of the country's energy history (2000–present), "energy crisis." China's energy demand began soaring again, rising much more rapidly than GDP. Recently, the country has been adding the equivalent of California's entire generating capacity every year. Most of the new power is from traditional coal plants, none of which can be easily retrofitted to capture carbon dioxide. As of 2005, China was burning twice as much coal as the United States. China now consumes more than twice as much steel as the United States and produces nearly as much cement as the rest of the world.

Oil demand has also been exploding, albeit beginning from a relatively low base. As of 2005, China still used less than one-third the oil that we do. And it has much higher fuel-economy standards than we do, as well as a 20 percent tax on the biggest gas-guzzling vehicles. But China has an exploding middle class, its passenger-car market increased tenfold in the 1990s, and it has been adding highways so fast that their total length is now second only to that of the United States. Worse still, China is pursuing several coal-to-diesel demonstration projects, and plans to replace 10 percent of projected oil imports in 2020 with that most carbon-intensive of liquid fossil fuels.

A 2005 study by the National Center for Atmospheric Research looked at our large and growing trade deficit with China. The study found that if the United States had produced domestically all the products that it had imported from China, our emissions in 2003 would have been 6 percent higher and China's would have been 14 percent lower. Also, America's rate of growth in CO_2 emissions would have been nearly 50 percent higher from 1997 to 2003—which means we are exporting to China a huge fraction of our growth in greenhouse gas emissions. And since our manufacturing system is more efficient and less coal-intensive than China's, total global CO_2 emissions from 1997 to 2003 would have been lower by a stunning 720 million metric tons had we made the products we bought from China during that short period.

China, the United States, and the world are at a crossroads.

One path, the current path, leads to catastrophe. In 2004, China's carbon dioxide emissions rose an alarming 15 percent. If its recent emissions trend—and ours—continue unchecked, our two countries alone will be responsible for half of all growth in global carbon dioxide emissions from 2000 to 2025.

At a 2005 U.S.-China conference on coal sponsored by Harvard University, a senior Chinese official told me, "We hope your government will delay action" on climate change since "we benefit from your government policy." America's climate policy gives political cover to those in China who wish to continue their recent explosive growth in carbon emissions.

The Bush administration has not been content merely blocking domestic efforts to cut greenhouse gas emissions but has been actively trying to block international negotiations aimed at developing mandatory reduction targets beyond what Kyoto would require. Worse, the administration has been working hard to woo developing countries away from the UNFCCC Kyoto Protocol effort to develop global mandatory targets. It has launched the Orwellian-named Asia-Pacific Partnership on Clean Development and Climate, which rejects clean development. That partnership, whose members include the United States, Australia, China, India, Japan, and South Korea, explicitly rejects all mandatory efforts to reduce emissions, including caps.

Not surprisingly, the partnership endorses a strategy of voluntary action and technology development. It claims its strategy will reduce annual carbon emissions in 2050 from "reference case" levels of 22 billion tons down to 17 billion tons. But that "reference case" is the most extreme emissions trend line imagined by the Intergovernmental Panel on Climate Change. It represents a world with economic growth that is both very rapid and fossil fuel intensive. If carbon emissions in 2050 are 17 billion tons, we would be on the irreversible path to 80 feet of sea-level rise—even if there were no vicious cycles in the carbon system such as methane released from the melting tundra. *With* those powerful vicious cycles, we must keep global carbon emissions well below 10 billion tons in 2050.

The Asia-Pacific Partnership is a climate suicide pact. It is playing Russian roulette with six bullets in your gun.

America and the world must quickly jump off this path and onto a very different one. China must return to its strong efficiency efforts from the 1980s, while at the same time embracing a low-carbon strategy, including massive amounts of renewable energy and carbon capture and storage. The choking pollution in major Chinese cities, coupled with the energy bottlenecks and frequent blackouts found in most provinces, should be motivation enough— even ignoring the benefits of avoiding catastrophic sea-level rise and climate change that will devastate the country, with so much of its wealth along the coasts, with so much susceptibility to droughts and water shortages.

But as in our country, China's leaders operate under the misguided belief that they can pollute all they want during this time of rapid growth, then use their *future* wealth to solve their environmental problems. While that paradigm has worked in America for polluted rivers and smoggy cities, it is fatally flawed for dealing with the threat posed by irreversible climate impacts, such as the disintegration of the Greenland Ice Sheet or the release of the carbon and methane locked in the frozen tundra.

Most of the rest of the industrialized world is prepared to go down the only effective alternative path and has already made a baby step in the right direction by ratifying Kyoto. But as with restrictions on CFCs and the Montreal Protocol, the developing world will embrace the necessary mandatory restrictions on greenhouse gas emissions if and only if the United States leads the way forthrightly, and only if there is a broad-based strategy for the rich countries to help the poor countries embrace low-carbon development. So the next president of the United States must be a strong leader who makes climate the overriding priority.

In 2009, America must start with very strong domestic actions both to save ourselves and to send a clear signal to the rest of the world that we take moral responsibility for being by far the single biggest contributor to climate change. Second, we must then quickly bring together all the nations of the world to establish appropriate targets and timetables, ones that will distinguish between rich and poor countries, ones that keep atmospheric concentrations of carbon dioxide below 550 ppm. Any other course for this nation is self-destructive.

CHAPTER TEN

MISSING THE STORY OF THE CENTURY

In the end, adherence to the norm of balanced reporting leads to informationally biased coverage of global warming. This bias, hidden behind a veil of journalistic balance, creates . . . real political space for the US government to shirk responsibility and delay action regarding global warming.
——Maxwell Boykoff and Jules Boykoff, 2004

This is no time for men who oppose Senator McCarthy's methods to keep silent. We can deny our heritage and our history, but we cannot escape responsibility for the result.
——Edward R. Murrow, March 9, 1954

If we do not avert Hell and High Water, global warming will be the news Story of the Millennium. In a world where sea levels are rising a foot or more every decade for centuries, our coasts are ravaged by superstorms, and we face endless mega-droughts, global warming won't be the most important story—it will be the only story.

If we do avert catastrophe, global warming will still be the Story of the Century. Starting very soon, and for many decades to come, the top news will focus on the country coming together to embrace an aggressive government-led effort to preserve the American way of life by changing everything about how we use energy—on a scale that dwarfs what the nation achieved during World War II.

While the media has begun providing more coverage of global warming, that coverage is still a long way from adequately informing the public about the urgency of the problem and the huge effort needed to avert catastrophe. The media's miscoverage of global warming makes it much less likely that the country will act in time, and it is a key reason why only a third of Americans understand that global warming will "pose a serious threat to you or your way of life in your lifetime," according to a March 2006 Gallup Poll.

We don't have any Edward R. Murrows today, at least not on the climate issue. What we do have is a declining number of science reporters, and only a handful of those are dedicated to covering climate. Worse, the media has the misguided belief that the pursuit of "balance" is superior to the pursuit of truth—even in science journalism. The result is that global warming and its impacts are systematically underreported and misreported.

WHEN BALANCE ISN'T BALANCED

In November 2005, *Meet the Press* with Tim Russert held a remarkable discussion on the threat of avian bird flu. Russert began with a quote from Senate Majority Leader Bill Frist, a physician, who laid out an ominous scenario of "a fast-moving highly contagious disease that wipes out 5 percent of the world population," which the senator said had already happened once, in 1918. The Frist quote ends: "This glimpse into the past might be a preview to our future. An avian flu pandemic is no longer a question of if but a question of when."

Russert then spent a half hour discussing bird flu with Michael Leavitt, President Bush's secretary of Health and Human Services; Michael Ryan, director of the World Health Organization's Epidemic and Pandemic Alert and Response Department; Dr. Julie Gerberding, director of the Centers for Disease Control and Prevention; and Dr. Anthony Fauci, director of the National Institute of Allergy and Infectious Diseases. All four of these experts expressed great concern about avian bird flu and the urgent need for preemptive action.

Russert did not interview anyone who felt that the threat from bird flu had been exaggerated (and such experts do exist). He did not interview anyone who questioned the science of evolution, even though this bird flu can't become a pandemic unless the virus mutates to allow easy human-to-human transmission and even though the Bush administration itself has questioned the teaching of evolution in schools. As one evolutionary biologist wrote in 2005, "If we're unlucky, this virus will give us a nasty demonstration of evolution in action."

Russert asked Fauci how much of a possibility a pandemic flu really was and how worried should people be. Fauci, one of the country's most respected medical experts, pointed out that it wasn't a high-probability event, then added, *"But when you're dealing with preparing for something in which the consequences are unimaginable, you must assume, A, the worst-case scenario, and B, that it's going to happen"*(emphasis added).

That is precisely how we should think about global warming. The threat it poses to our nation and our planet is certainly as grave as that posed by avian flu, and potentially much more devastating. The consequences may be longer-term, but the time to start acting is equally short. And the scientific consensus about global warming is as strong as or stronger than that surrounding the possibility of a bird flu pandemic. Yet there has never been a *Meet the Press* devoted to global warming with four experts all warning the public of the looming danger and the urgent need for action.

I discussed the strong consensus on global warming in chapter 1. To repeat the key point, as *Science* editor in chief Donald Kennedy said back in 2001, "Consensus as strong as the one that has developed around this topic is rare in science." A 2004 analysis of nearly 1,000 peer-reviewed scientific studies concluded that "politicians, economists, journalists, and others may have the impression of confusion, disagreement, or discord among climate scientists, but that impression is incorrect."

This remarkable consensus creates a very large problem for the media when they choose to cover a scientific matter as a political debate and give equal time to "both sides." As long as a handful of U.S. scientists, most receiving funds from the fossil fuel industry, get equal time with hundreds of the world's leading climate scientists, the public inevitably ends up with a misimpression about the state of our scientific understanding. Nor can that ever change as long as the Denyers refuse to alter their views in the face of the evidence and the media keep refusing to weigh the evidence or present the consensus accurately.

This isn't real balance. It is the media putting its thumb on the scale.

Sadly, even the most respected newspapers fall into this trap, as seen in the study "Balance as Bias: Global Warming and the U.S. Prestige Press," which analyzed more than 600 hard-news articles published from 1990 to 2002 in the *New York Times, Washington Post, Los Angeles Times,* and *Wall Street Journal.* The study found that

- 53 percent of the articles gave roughly equal attention to the views that humans contribute to global warming and that climate change results exclusively from natural fluctuations
- 35 percent emphasized the role of humans while presenting both sides of the debate
- 6 percent emphasized doubts about the claim that human-caused global warming exists

- Only 6 percent emphasized the predominant scientific view that humans are contributing to Earth's temperature increases

The authors found a "significant difference between the scientific community discourse and the US prestige press discourse." As an example of balance as bias, consider these lines from an April 2001 *Los Angeles Times* article:

> The issue of climate change has been a topic of intense scientific and political debate for the past decade. Today, there is agreement that the Earth's air and oceans are warming, but disagreement over whether that warming is the result of natural cycles, such as those that regulate the planet's periodic ice ages, or caused by industrial pollutants from automobiles and smokestacks.

Notice how science and politics become merged, and the reader is left with the distinct impression that there is an intense scientific disagreement about whether the warming has a natural or a human-made cause. But there is no such disagreement. Few climate scientists doubt that most of the warming is human-caused and, equally important, that human-caused warming will increasingly dwarf all natural trends.

The media's pursuit of "balance," coupled with their growing desire for drama and entertainment, has left them vulnerable to targeted campaigns of misinformation. To create doubt on any scientific issue, all you have to do is find a few credible-sounding people to present your side, and no matter how many people are on the other side, you've got instant debate. This exploitable flaw in the coverage of science has not gone unnoticed by the global-warming Delayers. As the *New York Times* reported in April 1998, the fossil fuel industry developed a draft plan "to spend millions of dollars to convince the public that the [Kyoto] environmental accord is based on shaky science." Its major strategy was "a campaign to recruit a cadre of scientists who share the industry's views of climate science and to train them in public relations so they can help convince journalists, politicians and the public that the risk of global warming is too uncertain to justify controls on greenhouse gases like carbon dioxide."

The amount of media coverage of global warming has improved in the last few years, likely because the weight of scientific evidence plus the consensus about the dangers of inaction have become too strong to ignore. Yet most articles on climate are still confusing or misleading or both. Let's look at a few 2006 articles from the *Washington Post*, a newspaper that has done some of the media's best reporting on global warming.

Consider a short January 23 article on a *Nature* paper that "suggests that melting mountain glaciers and ice caps, which account for about a quarter of the expected sea level rise, will produce about half the level of sea level rise by 2100 others have predicted." You might expect the article would be balanced with an expert explaining why scientists are far more concerned with observations of accelerated disintegration of the Greenland and Antartic ice sheets, which contain far more ice and which this study doesn't examine at all. Instead, the article quotes Pat Michaels, of the Marshall and Cato Institutes, both funded by ExxonMobil to advance the Denyers' agenda.

Michaels is quoted saying the *Nature* paper "is one of many recent papers pointing towards reductions in sea level rise in this century due to more refined models of ice balance"—a claim that is best described as the opposite of the truth. Indeed, six days later, on January 29, the *Post* itself got the story straight and published a front-page article noting, "Most scientists agree human activity is causing Earth to warm," so "the central debate has shifted to whether climate change is progressing so rapidly that, within decades, humans may be helpless to slow or reverse" key impacts such as "dramatic sea level rise by the end of the century that would take tens of thousands of years to reverse."

A July 2006 coal-industry memo revealed how the industry is funding Michaels as part of its strategy to stop action on global warming. The Associated Press led its story, "Coal-burning utilities are passing the hat for one of the few remaining scientists skeptical of the global warming harm caused by industries that burn fossil fuels." That article also explained how Michaels misrepresented James Hansen's testimony in an effort to discredit him (see chapter 5).

Consider a May 3, 2006, *Washington Post* article on how the new conservative government in Canada is cutting programs to reduce greenhouse gas emissions. The article explained that in the Kyoto Protocol, countries "pledged to meet quotas to reduce the carbon dioxide emissions that many scientists believe are warming Earth, melting glaciers and brewing more intense storms." Such misleading sentences serve only to confuse the public. The overwhelming majority of scientists believe carbon dioxide emissions are warming the earth and melting glaciers, as the earlier January 29 *Post* article had noted. And the scientific literature is clear that global warming makes storms more intense; the debate on this issue is primarily over *how much* more intense.

The article balances quotes from Canadians who believe the country should take action on climate change with quotes from Morten Paulsen of Friends of Science, a group of Delayers and Denyers with links to the fossil fuel industry. According to Paulsen, "We shouldn't be spending billions of dollars fighting a problem that may not be there." The article states, "He said that arguments that global warming is caused by carbon dioxide are unproven and that 'we believe they are a white elephant.'"

Arguments that global warming is caused by carbon dioxide are *not* unproven. Countless studies have been published on this, all major scientific bodies that have looked at the question acknowledge this as a fact, and it would be hard to find 1 scientist in 10,000 who would agree with Paulsen's claim. Would the *Post* quote someone denying that we had landed on the moon? Would the *Post* quote a tobacco-company lobbyist saying, "Arguments that cancer is caused by cigarette are unproven"?

Consider another *Washington Post* article from the same day, May 3, on a major government study that "undermines one of the key arguments of climate change skeptics, concluding that there is no statistically significant conflict between measures of global warming on the earth's surface and in the atmosphere." For more than a decade the Denyers have argued that global warming could not be happening because the measured warming of the earth's surface was apparently not matched by the satellite measurements of the atmosphere's temperature—measurements first analyzed and reported by University of Alabama researchers led by John Christy. Christy's analysis had suggested a temperature *decrease* in the satellite data. As one encyclopedia notes, however, other scientists "over the years have shown errors in his interpretation of the data which has slowly and consistently increased his results."

Christy, like Michaels, is among a handful of scientists regularly quoted by the media for "balance." While the number of scientists reporting evidence of human-induced climate change multiplies with each passing year, the "balancers" remain a group small enough to fit into a typical home bathroom. Or even its shower. Christy contributed to a 2002 book called *Global Warming and Other Eco-Myths,* published by the Competitive Enterprise Institute, which is funded by ExxonMobil.

Science magazine begins its article on that same 2006 government study: "Global warming contrarians can cross out one of their last talking points." *Science*'s headline trumpets the news: "No Doubt About It, the World Is Warming." Such a stunning vindication for climate scientists needs no quote from Denyers for phony balance. The *Post*, however, spends nearly half the article quoting James Inhofe and John Christy dismissing the relevance of the blockbuster report. Inhofe's spokesman repeats the discredited natural-cycles argument, which the *Post* article does not rebut. Christy claims the earth isn't heating up rapidly enough for him to be very worried, an assertion the article also chooses not to rebut. So an article that should read as a crushing blow to global-warming Denyers instead becomes a vehicle for them to rehash dubious and discredited arguments, with little or no check by the newspaper.

The *Science* article isn't quite perfect. It says the new report, though commissioned by the Bush administration, "will not change White House policy." It then paraphrases a White House spokesperson: "President George W. Bush believes that greenhouse gas emissions can be brought down through better use of energy while the understanding of climate science continues to improve." If Bush really believes that, he has never publicly stated it, nor has he pursued a single policy to achieve reductions in emissions through better use of energy. The spokesperson, or the reporter, may have been confused or mistaken—or meant that Bush believes greenhouse gas emissions *intensity* (per unit GDP) can be brought down through better use of energy. Either way, someone reading the article would be left with the mistaken impression that Bush is actually pursuing energy strategies that reduce emissions.

I can't see why serous news outlets would quote Michaels or Christy on climate science. Those that do quote Michaels should follow AP's lead in explaining that he has been intentionally misleading and is heavily subsidized by the coal industry. Those that quote Christy should explain how he consistently misanalyzed key data and then trumpeted his mistaken conclusions as proof that global warming wasn't happening, long after other scientists explained that he was wrong.

Then there is meteorologist Bill Gray, who testified at a 2005

Senate hearing that we will be headed back into a period of global cooling in a few years and that climate science is just a hoax created by the scientific community to get more funding. Gray is typically described as a great hurricane forecaster, as in a 2006 *Washington Post Magazine* cover story. You would never know from such coverage that shortly after the 2004 hurricane season, he predicted, "We probably won't see another season like this for a hundred years." He was off by only 99 years.

How consistently wrong do you have to be before the media stops quoting you as an expert?

"If your mother says she loves you, check it out" was the adage journalists like my father were schooled on. Be skeptical of even the most obvious truths and check your facts, yes, but nowhere does the motto say to ignore the truth or assume there is none. Today the media's motto seems to be "If your mother says she loves you, get a quote from the neighborhood bully."

EVERYBODY TALKS ABOUT THE WEATHER

One area of media miscoverage in this country deserves particular mention. The key message about what is happening has been muffled. That message is: Climate change is a driving force behind the increasing amount of extreme weather we are experiencing.

Consider a *New York Times* article from July 2003, "Records Fall as Phoenix All but Redefines the Heat Wave," highlighting daytime temperatures of 117°F and nighttime temperatures of 96°F—"the hottest night in Phoenix history." The article never suggests even the possibility that global warming has contributed to redefining the heat wave or that scientists expect such heat waves to become not only more commonplace but more severe.

Consider a *Washington Post* article from the same month, "Coastal Louisiana Drowning in Gulf: Encroaching Salt Water Is

Threatening the State's Economy and Homes." The article discusses a variety of reasons Louisiana annually loses more than 25 square miles of coastland to the Gulf of Mexico, such as efforts to control the flow of the Mississippi River. Nowhere does the article mention even the possibility that climate change has contributed to the problem or that future sea-level rise, left unchecked, may undermine all efforts to find a long-term solution.

Consider a January 2006 NBC News report on extreme weather titled "Meltdown." The report starts in New York, which in midwinter was experiencing springlike weather with temperature in the 50s. It shows reporter Mike Taibbi hitting golf balls in a short-sleeve shirt and getting advice from a golfing pro. After jumping to footage of unusual weather around the country and the world, Taibbi talks to NBC meteorologist Jeff Ranieri:

> **TAIBBI**: But the unseasonable weather isn't restricted to the Northeast. With twenty-five straight days of downpour, Seattle and the Pacific Northwest are approaching rainfall records. Extreme heat and lack of rain have fed the wildfires tormenting parts of Oklahoma and Texas. Rare ocean tornadoes have been seen off the Florida coast. And in usually frigid Chicago, kids eating ice cream cones watch flamingos and giraffes take the sun. . . . Around the world, more extreme weather; the snowiest winters in generations in parts of Japan and China. The cause of all this?
>
> **RANIERI**: *I wouldn't say that this is, uh, a long-term pattern that we're stuck in. It's just . . . it's Mother Nature* and it's just how it's working in the beginning of January.
>
> **TAIBBI**: Back to the *thoroughly enjoyable extreme weather* in New York . . . [Emphasis added.]

Wrong, wrong, and wrong. As the chapter 2 discussion of the U.S. Climate Extremes Index makes clear, it *is* a long-term pattern.

The pattern is *not* what we expect from Mother Nature, but it is precisely what we *do* expect from global warming. And while it may be enjoyable in wintertime for New Yorkers, it is catastrophic for those suffering from flooding and wildfires.

If the media's coverage of weather extremes does not improve in the next few years, we will have no chance of avoiding the disintegration of the great ice sheets. Ironically—and as we have seen throughout the book, irony is the defining characteristic of the global-warming debate—the only truly prophetic element of the NBC story was its title, "Meltdown."

Such bad coverage has consequences: Even sophisticated people are left uninformed. Consider Lisa Murkowski, Republican senator from Alaska, the state most strongly hit by the effects of climate change, who sits on the Senate Environment Committee and casts votes that determine the nation's climate policy. Near the end of a September 2005 hearing on climate science, she pointed out that Alaska had experienced "continuous erosion of our coastal villages" and the "warmest summer that we've seen in 400 years." What does she think of all this? "I'm sitting up in Alaska where I can see that we're experiencing climate change. *I'm not going so far as to say it is global warming. But we see climate change*" (emphasis added).

Why take any serious action now if it all might just be a natural event, purely a coincidence that it is occurring at the same time that we're putting into the atmosphere massive amounts of greenhouse gases that scientists predicted would cause these exact changes?

How did the media coverage get so bad? The story should be as simple and logical as the story about avian flu. We have an overwhelming consensus among our leading scientists that global warming is happening and humans are the primary cause. We know that one of the earliest expected impacts of global warming is an increase in extreme weather events. We have a painfully obvious increase in extreme weather.

We even have the federal agency in charge of climatic data, the

National Climatic Data Center, with a comprehensive statistical measure showing that the weather has actually gotten more extreme—and which explained more than ten years ago that the chances that this increase was due to factors *other* than global warming, such as "natural climate variability," was statistically very small.

Yet my guess is that you've never heard of the U.S. Climate Extremes Index, even though it was explicitly created to take a complicated subject ("multivariate and multidimensional climate changes in the United States") and make it more easily understood by American citizens and policy makers. I follow this subject of the connection between climate change and extreme weather very closely, and yet, until 2006, I had not seen a single mention of the index in the media or even in a scientific paper since its original introduction back in 1995.

Story after story in the media appear with no link whatsoever between extreme weather and global warming, no link to the human-made trend that will ultimately transform all our lives. Even the monster U.S. heat wave at the end of July and early August 2006 generated few stories that mentioned global warming. I was actually interviewed by a major national news outlet about this heat wave. They were interested in my work on urban heat islands, whereby dark roofs and asphalt pavement and the loss of shade trees have made cities much hotter than they would otherwise be (see chapter 6). Although I discussed how global warming is making this kind of devastating heat wave more likely and more intense—and combining with the heat-island effect to make cities increasingly inhospitable in the summer—they did not use any of these comments. They wanted only a story on how heat islands affect heat waves.

What are the reasons for this flawed and incomplete reporting?

One reason is that the Delayers have been hard at work criticizing the media for making the link between extreme weather and climate change—and they've succeeded in intimidating them. In his 2004 book, *Boiling Point,* Pulitzer Prize–winning journalist Ross Gelbspan wonders why journalists covering extreme weather events don't use the statement "Scientists associate this pattern of violent weather with global warming." He reports that a few years earlier he had asked "a top editor at a major TV network" why they didn't make this link. The reply was: "We did that. Once. But it triggered a barrage of complaints from the Global Climate Coalition [then the major anti-global-warming lobbying group of the fossil fuel industry] to our top executives at the network."

The lobbyists argued then, as they do now, that you can't prove that any individual weather event is caused by climate change. But that is irrelevant to the two key points: The pattern is exactly what we expect from climate change, and we can expect to see more violent weather events in the future if emissions trends are not reversed soon.

Another reason the media gets the climate extreme-weather link wrong: Most meteorologists in this country, including virtually every TV meteorologist, are not experts on global warming. As one climate scientist explained to me:

> Meteorologists are not required to take a course in climate change, this is not part of the NOAA/NWS [National Oceanic and Atmospheric Administration/National Weather Service] certification requirements, so university programs don't require the course (even if they offer it). So *we have been educating generations of meteorologists who know nothing at all about climate change.* [Emphasis added.]

Asking a meteorologist to explain the cause of recent extreme weather is like asking your family doctor what the chances are for an avian flu pandemic in the next few years or asking a Midwest sheriff about the prospects of nuclear terrorism. The answer might be interesting, but it wouldn't be one I'd stake my family's life on.

A final reason you don't see the link made here in this country as much as you should is that the environmental community itself decided in the mid-1990s to *deemphasize* it. Yes, you read that right. Many environmentalists actually made a conscious decision to stop talking about what are arguably the most visible and visceral signs of warming for most people. A number of senior environmentalists, including those involved with media outreach, told me at the time that they were tired of being beaten up by the other side on this issue. I thought that was a blunder then, and I still do today.

Peter Teague, Environment Program director for the Nathan Cummings Foundation, wrote about this problem in the summer of 2004 after "the fourth in a series of violent hurricanes [had] just bombarded the Caribbean and Florida." He pointed out "no prominent national leader—environmental or otherwise—has come out publicly to suggest that the recent spate of hurricanes was the result of global warming."

But the ever-worsening reality of climate change together with the diligence of leading climate scientists brought the hurricane-warming link roaring back. As noted in chapter 2, leading scientists from MIT, the Georgia Institute of Technology, and the National Center for Atmospheric Research (NCAR) published a series of scientific articles on the rise of intense hurricanes—in what turned out to be the most devastating hurricane season in U.S. history, 2005. And in the months following Katrina, the scientific basis for the connection between global warming and intense hurricanes has grown even stronger.

The media still do not cover the story well. In a major article on climate change in April 2006, the *New York Times* actually claimed, "Few scientists agree with the idea that the recent spate of potent hurricanes, European heat waves, African drought and other weather extremes are, in essence, our fault." Few? That doesn't gibe with the dozens of climate scientists I talked to while researching this book. They all told me what climate scientists have been telling us so many times before—global warming makes extreme weather events more likely and more destructive.

Again, the story is fairly straightforward. Global tropical sea-surface temperature is increasing as a result of greenhouse warming. Average hurricane intensity increases with increasing tropical sea-surface temperature. The frequency of the most intense hurricanes is increasing globally. So greenhouse warming is causing an increase in global hurricane intensity. True, not every scientist agrees with that conclusion, but fewer and fewer are disagreeing, while more and more are speaking out bluntly. "The hurricanes we are seeing are indeed a direct result of climate change, and it's no longer something we'll see in the future, it's happening now," Greg Holland, an NCAR division director, told the American Meteorological Society's 27th Conference on Hurricanes and Tropical Meteorology in April 2006.

That is what I told my brother, to aid in his decision about whether to rebuild or relocate from the Gulf Coast. That is what everyone making such decisions needs to hear to make an informed choice.

THE STORY OF THE CENTURY: BE *VERY* WORRIED

Most of the media do not get global warming—yet. And that extends from TV and radio to newspaper to magazines to even the most sophisticated policy journals such as *Foreign Affairs,* which routinely publishes major articles on subjects like China and energy with virtually no mention of global warming. One publication, however, has consistently delivered timely and powerful stories on global warming, largely unfettered by faux balance—*Time* magazine.

In April 2006, *Time* published a powerful special report on global warming with a warning on the cover in huge letters, "BE WORRIED. BE *VERY* WORRIED. Climate change isn't some vague future problem—it's already damaging the planet at an alarming pace." One of the most interesting things in the issue was a poll in which 1,000 Americans were asked, "Do you think most scientists agree with one another about global warming, or do you think there is a lot of disagreement on this issue?" Only 35 percent said, "Most agree," while 64 percent said, "A lot of disagreement." As *Time* noted, "Most people aren't aware of the broad scientific consensus on warming." But then how could they be, with other media continually misreporting the subject, insisting that as long as there is one global-warming Denyer, that Denyer deserves equal time with the entire rest of the scientific community?

In a fascinating example of intramedia "balance," *Time*'s rival, *Newsweek,* also published an article on global warming that week. Unlike *Time, Newsweek* devoted almost half of its article to quoting various Denyers and Delayers, claiming, "To be fair, neither side has a monopoly on hot air in this debate," falsely equating one or two mild overstatements by advocates of action on global warming with major campaigns to deny the science entirely and delay action indefinitely.

The *Newsweek* article seeks to downplay the growing concern over warming: "But both the [Elizabeth] Kolbert and [Tim] Flannery books are sober, detailed and alarming without being alarmist." Yet Kolbert's book is titled *Field Notes from a Catastrophe,* and the final sentence is "It may seem impossible to imagine that a technologically advanced society could choose, in essence, to destroy itself, but that is what we are now in the process of doing." Flannery's book warns that if we don't act fast enough to limit greenhouse gas emissions, we will "destroy Earth's life-support systems and destabilise our global civilisation." The result: "Humans are thrust into a projected Dark Ages far more mordant than any that has gone before. . . . These changes could commence as soon as 2050."

Both Kolbert's and Flannery's books strike me as alarming *and* alarmist—as befits any sober and detailed examination of the facts.

The subhead on the *Newsweek* article is "Books, Films and a Slick Ad Campaign Make Global Warming the Topic *Du Jour*." No. Global warming is the topic *du siècle*. And if we don't get a lot more stories, and a lot better stories, on the threat and how to stop it, global warming will be the only story that matters to the next fifty generations of Americans.

CONCLUSION

THE END OF POLITICS

The hottest places in Hell are reserved for those who in time of great moral crises maintain their neutrality.
—attributed to Dante

America is great because she is good. If America ceases to be good, America will cease to be great.
—attributed to Alexis de Tocqueville

Global warming will change American life forever and end politics as we know it, probably within your lifetime. How might this play out?

In the best case, we immediately start changing how we use energy in order to preserve the health and well-being—the security—of the next fifty generations. The nation and the world embrace an aggressive multidecade, government-led effort to use existing and near-term clean-energy technologies.

The enabling strategy is energy efficiency—since that generates the savings that pays for the zero-carbon energy sources, like wind power and coal with carbon sequestration. Efficiency keeps the total cost low to consumers and businesses. For utilities, we need a California-style energy-efficiency effort nationwide. For cars and light trucks, we need serious federal standards for high-mileage hybrids that can be plugged in to the electric grid. The goal of all these efforts: keeping global emissions at or below 29 billion metric tons of carbon dioxide (8 billion tons of carbon) for the next several decades—and keeping concentrations well below 550 ppm (a doubling of preindustrial levels) this century.

I have called this scenario Two Political Miracles because it would require a radical conversion of American conservative leaders—first, to completely accept climate science, and second, to strongly embrace climate solutions that they currently view as anathema. I have spent nearly two decades working to achieve this clean-energy future and will continue doing so, because it is the best way to preserve the health and well-being of future generations and to boost energy security while creating millions of clean-energy jobs here at home. Yet none of the more than one hundred people I interviewed for this book considers this in the least bit plausible.

They may be right. Tragically, in the face of the stunning recent evidence that climate change is coming faster and rougher than scientists have expected, many conservatives have chosen to redouble their efforts to deny the science and delay serious action. Consider the words of President Bush in May 2006: "In my judgment, we need to set aside whether or not greenhouse gases have been caused by mankind or because of natural effects." That statement is reminiscent of leaders like Herbert Hoover and Neville Chamberlain who were blind to their nation's gravest threats.

President Bush misspoke. The massive surge in greenhouse gas emissions is clearly caused by humankind—that is not even in dispute. What Bush may have meant to say is "climate change" rather than "greenhouse gases," which is the standard rehashing of the long-discredited "climate change is all just natural cycles" argument. We cannot, however, set aside the overwhelming evidence and solid scientific consensus that humankind is to blame for virtually all of recent climate change because that would mean setting aside the possibility of any serious effort to prevent future catastrophic climate change from human emissions.

Consider two ads launched in May 2006 by the Competitive Enterprise Institute, an oil-industry-funded think tank. One claims that the Greenland and Antarctic ice sheets are increasing in mass due to increased snowfall. The ad conveniently ignores the evidence that both ice sheets are now losing ice at the edges faster than they are gaining mass in the center—and doing so much faster than predicted. As recently as 2001, the international scientific community thought that the great ice sheets would not contribute significantly to sea-level rise this century. But as climatologist Richard Alley warned, also in May 2006, "The ice sheets seem to be shrinking 100 years ahead of schedule."

Both ads end with a rhetorical tagline that would be funny if the stakes weren't so deadly serious: "Carbon dioxide—they call it pollution, we call it Life!" Yes, carbon dioxide is needed for life, as is water. But too much of either can be fatal. Just look at New Orleans and the Gulf Coast. The Competitive Enterprise Institute might just as well have ended its ads, "Après nous le deluge" (*After us, the deluge*)—literally. Under the Competitive Enterprise Institute's banner, we would never take any action whatsoever to reduce carbon dioxide emissions, even to avoid a tripling or quadrupling of preindustrial concentrations. Does the conservative movement really want to side with global-warming pollution over the health and well-being of the next fifty generations?

The conservative Denyers and Delayers are not the only reason America has failed to take up the fight against climate change. "Scientists present the facts about climate change clinically, failing to stress that business-as-usual will transform the planet," leading to as much as 80-feet-higher sea levels, rising "twenty feet or more per century," as NASA's Jim Hansen wrote in 2006.

Progressive politicians have been slow to grasp the overwhelming urgency of the problem. But that is starting to change. Al Gore has launched a major effort to mobilize action, built around his 2006 movie, *An Inconvenient Truth*. Also in 2006, Democrats in both the House and Senate have for the first time introduced legislation that would require reductions in greenhouse gas emissions sufficient to avert catastrophe.

Some major groups that have been on the sidelines, such as evangelical Christians, have begun reconsidering their position on climate. In February 2006, the Evangelical Climate Initiative, a group of more than 85 evangelicals, issued a statement saying, "Human-induced climate change is real," the "consequences of climate change will be significant," and government should immediately pass legislation reducing U.S. carbon dioxide emissions. In response, however, key conservative evangelicals launched the Interfaith Stewardship Alliance, "which has aligned itself with prominent global warming skeptics, including John Christy and . . . Patrick Michaels," as an April 2006 news story explained. In July 2006, the Heritage Foundation hosted an event at which the alliance released a letter, signed by more than 100 evangelicals, questioning the science of climate change; the letter claimed that global warming will have mild and possibly "helpful" consequences and opposed any "government-mandated carbon dioxide emissions reductions," saying they "would cause greater harm than good to humanity."

So I do believe that for all of the failings of the groups seeking strong action on global warming, the conservative Denyers and Delayers are the main reason America lacks the consensus and the political will to take up the fight against catastrophic climate change. They actively spread misinformation. They block those seeking to take action at a state, national, and global level. We will not be able to prevent catastrophic warming without conservatives embracing a dramatically different view of energy policy, international negotiations, and the role of government.

For now, the political success of the global warming Delayers must be acknowledged. No proposal to reduce U.S. greenhouse gas emissions has ever achieved a majority vote in either chamber of Congress. America will almost certainly take no serious action on climate under President Bush, and he may very well block any serious efforts by other nations. Long after Bush leaves office, conservatives in Congress will hold enough strength to block significant action on climate, should they so choose. This suggests that America will at best take some half measures to slow our emissions growth in the next decade, while China and other developing nations continue their breathtaking emissions growth largely unchecked. That will put us on the brink of disaster.

THE RECKONING

Soils, tundra, tropical forests, and oceans currently serve as sinks that absorb nearly half the carbon we are spewing into the atmosphere. The tundra by itself today contains about as much carbon as the atmosphere, much of it in the form of methane, which is more than twenty times as potent at trapping heat as carbon dioxide. At 550 parts per million of atmospheric carbon dioxide concentrations, a doubling of preindustrial levels, we are likely to lose most of the tundra and most of the Amazon rain forest, and with them any hope of avoiding a tripling, which would ruin this planet for the next fifty generations.

But barring the Two Political Miracles, global emissions will hit 37 billion metric tons of carbon dioxide a year in the early 2020s, while global concentrations hit about 430 ppm, rising 3 ppm a year. We will have vastly overshot a safe level of carbon emissions, and misallocated trillions of dollars in capital constructing conventional coal plants, producing unconventional oil, and manufacturing inefficient vehicles. At that point, if we wanted to avoid climate catastrophe while avoiding economic collapse, we would have no choice but to scrap most of this polluting capital long before the end of its natural life, while replacing it with clean, efficient capital at a rapid rate.

This national (and global) reindustrialization effort would be on the scale of what we did during World War II, except it would last far longer. "In nine months, the entire capacity of the prolific automobile industry had been converted to the production of tanks, guns, planes, and bombs," explains Doris Kearns Goodwin in her 1994 book on the World War II home front, *No Ordinary Time.* "The industry that once built 4 million cars a year was now building three fourths of the nation's aircraft engines, one half of all tanks, and one third of all machine guns."

The scale of the war effort was astonishing. The physicist Edward Teller tells the story of how Niels Bohr had insisted in 1939 that making a nuclear bomb would take an enormous national effort, one without any precedent. When Bohr came to see the huge Los Alamos facility years later, he said to Teller, "You see, I told you it couldn't be done without turning the whole country into a factory. You have done just that." And we did it all in less than 5 years.

But of course we had been attacked at Pearl Harbor, the world was at war, and the entire country was united against a common enemy. This made possible tax increases, rationing of items such as tires and gasoline, comprehensive wage and price controls, a War Production Board with broad powers (it could mandate what clothing could be made for civilians), and a Controlled Material Plan that set allotments of critical materials (steel, copper, and aluminum) for different contractors.

Such desperate and undesirable national actions are a long, long way from mandated controls on carbon dioxide emissions or requiring that 20 percent of all power come from renewable sources—neither of which conservatives currently support. The ultimate irony would be if conservative disdain for straightforward government-led solutions *today* forced the country into far more intrusive and onerous government solutions *tomorrow*.

And what happens if the nation and the world fail to take serious action in the 2020s? In the 2030s, record-breaking heat waves and searing droughts will be the norm. Relentless super-hurricane seasons, coupled with the reality of accelerating sea-level rise, will change the landscape of the Gulf Coast and the eastern seaboard. We will simply stop rebuilding most coastal cities destroyed by hurricanes. In this Planetary Purgatory, everyone will realize that the world has but one great task—stopping Greenland and the West Antarctic Ice Sheet from melting, avoiding runaway growth in greenhouse gas concentrations.

Politics as we know it will end. Nonessential efforts, such as the manned space program, will be shut down as politicians direct most of the nation's vast resources toward dealing with the climate. The problem with waiting until the 2030s is that carbon dioxide concentrations are likely to be over 450 ppm and climbing more than 3 ppm a year. At that point, our fate will be largely out of our hands and in the hands of the vicious carbon cycles. Most likely we will be headed irrevocably toward Hell and High Water—a tripling of concentrations or worse, warming of the inland United States of 10°F or worse, sea-level rise exceeding 1 foot a decade, widespread ecosystem collapse, and mass extinctions.

The suffering that my brother and his family and the hundreds of thousands of victims of Hurricane Katrina experienced will be magnified a thousandfold in a world with half a billion environmental refugees, water and food shortages affecting a billion or more people, and worldwide civil strife.

We must pay any price and bear any burden to avoid this fate.

What would have to happen in the next decade to create the political will needed to transform the entire country into a carbon-reducing factory? I see two possibilities. The first requires that a major climatic event or series of events occur. A portion of the West Antarctic Ice Sheet could disintegrate rapidly, raising sea levels 20 inches. Or the country could be hit by the kind of murderous heat wave that overwhelmed Europe in 2003. Or we could experience several more hurricane seasons like 2005. Or, more likely, all of those, since the national and global heat wave of 2006 does not appear to be changing U.S. climate politics.

Second, the public—you—could simply demand change. This is vastly preferable to waiting for multiple disasters. Global warming is the gravest threat to our long-term security. More and more people are coming to this realization every day. When people ask me what they should do, I reply, "Get informed, get outraged, and then get political." I think it is a good idea to take steps to reduce your own greenhouse gas emissions, purchasing a hybrid vehicle, buying Energy Star home appliances, buying renewable power, encouraging your workplace to take action—mainly so that you can see that taking action is not that hard.

You must become a climate champion, a single-issue voter. You must take whatever action you can. You must use whatever influence you have wherever it would make a difference, even if it is only to educate the people around you.

I do believe that if we fail to act in time, it will be the single biggest regret any of us has at the end of our lives.

So you can see why my hair *is* on fire. I hope yours is, too.

ACKNOWLEDGMENTS

While the final judgments in the book are my own, I am exceedingly grateful to everyone who shared their ideas with me: Richard Alley, John Atcheson, Peter Barrett, Bill Becker, Dan Becker, Alec Brooks, Brian Castelli, Joseph Cirincione, Ana Unruh Cohen, Jon Coifman, Bob Corell, Judith Curry, Noah S. Diffenbaugh, Kerry Emanuel, Alex Farrell, Tim Flannery, Louis Fortier, Peter Fox-Penner, Andrew Frank, S. Julio Friedmann, Tore Furevik, David Gardiner, Dean Grodzins, Jason Grumet, Hank Habicht, David Hamilton, Kate Hampton, James Hansen, David Hawkins, Bracken Hendricks, John Holdren, Roland Hwang, Andrew Jones, Chris Jones, Tina Kaarsberg, Dan Kammen, Thomas Karl, David Keith, Henry Kelly, Melanie Kenderdine, Felix Kramer, Kalee Kreider, David Lawrence, Mark Levine, Lee Lynd, Jerry Mahlman, Wieslaw Maslowski, Jan Mazurek, Alden Meyer, Ron Minsk, Ernest Moniz, Philip Mote, Walter Munk, Joan Ogden, Michael Oppenheimer, Jonathan Overpeck, John Passacantando, John Podesta, Arthur Rosenfeld, Doug Rotman, Gavin Schmidt, Stephen Schneider, Dan Schrag, Laurence Smith, Robert Socolow, Kevin Trenberth, Peter Webster, Robert Williams, James Woolsey, Paul Wuebben, and Jay Zwally.

I am grateful to James Fallows for suggesting that I write this book. Special thanks go to those who reviewed all or part of it: Ana Unruh Cohen, Judith Curry, Jay Gulledge, Greg Kats, and Steve Silberstein. I would particularly like to thank Pete O'Connor for his help on research and references, and John Atcheson for his help and support. I would like to thank the Hewlett Foundation for its support of my work.

I owe a permanent debt to my mother for applying her unmatched language and editing skills to innumerable drafts. I am exceedingly grateful to Peter Matson, my agent, for his unwavering efforts on behalf of my work and my writing. He found this book the perfect home.

Henry Ferris, my world-class editor at William Morrow, improved the book tremendously. I would like to thank him and the rest of the staff at William Morrow for their enthusiastic support of this book from the very beginning.

Finally, special thanks go to my wife, Patricia Sinicropi, whose unconditional love and support provide me unlimited inspiration every day. Words cannot express what a remarkable person she is.

NOTES

INTRODUCTION

1 *"We are on":* James Hansen, "Is There Still Time to Avoid 'Dangerous Anthropogenic Interference' with Global Climate?" (paper presented at the American Geophysical Union, San Francisco, December 6, 2005).

1 *"The ice sheets":* Richard Alley, American Meteorological Society seminar, Washington, D.C., May 3, 2006.

2 *March 2006 Gallup Poll:* Gallup Poll, March 13–16, 2006, available online at http://www.pollingreport.com/enviro.htm.

5 *advanced acoustic techniques:* Joseph Romm, "Applications of Normal Mode Analysis to Ocean Acoustic Tomography" (Ph.D. diss., Massachusetts Institute of Technology, researched at the Scripps Institution of Oceanography, 1987).

6 *"straw that breaks":* Kevin Trenberth (American Meteorological Society seminar, Washington, D.C., October 25, 2005).

7 *National Academy of Sciences:* Walter Sullivan, "Scientists Fear Heavy Use of Coal May Bring Adverse Shift in Climate," *New York Times,* July 25, 1977.

CHAPTER ONE: THE CLIMATE BEAST

11 *"The paleoclimate record":* Wallace S. Broecker, "Cooling the Tropics," *Nature* 376, no. 6357 (July 20, 1995): 212–13.

11 *"The ongoing Arctic":* Louis Fortier, "The Arctic as a Bellwether for Climate Change" (paper presented at the Arctic Climate Symposium, Washington, D.C., June 15, 2006).

11 *"hypertext history":* American Institute of Physics website, www.aip.org/history/climate/.

12 *"the stated degree":* National Research Council, *Climate Change Science: An Analysis of Some Key Questions* (Washington, D.C.: National Academies Press, 2001).

12 "Consensus as strong": Donald Kennedy, "An Unfortunate U-Turn on Carbon," *Science* 291, no. 5513 (March 20, 2001): 2515. See also S. Fred Singer, "Editor Bias on Climate Change?" and Donald Kennedy's response, *Science* 301, no. 5633 (August 1, 2003): 595–96.

12 Science *published:* Naomi Oreskes, "Beyond the Ivory Tower: The Scientific Consensus on Climate Change," *Science* 306, no. 5702 (December 3, 2004): 1686.

13 *"Evidence of global":* Robert Kunzig, "Turning Point," *Discover,* January 2005: 26–28.

13 *"There can no longer":* James Hansen, "Answers About the Earth's Energy Imbalance," Earth Institute at Columbia University, 2005, available online at www.earthinstitute.columbia.edu/news/2005/story11-04-05_html; James Hansen et al., "Earth's Energy Imbalance: Confirmation and Implications," *Science* 308, no. 5727 (June 3, 2005): 1431–35.

13 *a joint statement:* "Joint Science Academies' Statement: Global Response to Climate Change," June 7, 2005, available online at http://nationalacademies.org/onpi/06072005.pdf.

14 *heat, infrared radiation:* Wikipedia, s.v. "greenhouse effect," http://en.wikipedia.org/wiki/Greenhouse_effect.

15 *ideal for us humans:* Glass greenhouses achieve most of their enhanced warming by physically stopping hot air from leaving the enclosure. The atmospheric greenhouse effect is thus a different type of process, since it relies on gases such as carbon dioxide trapping infrared radiation.

16 *Figure 1:* Jean-Rubert Petit et al., "Historical Isotropic Temperature Record from the Vostok Ice Core," January 2000, Carbon Dioxide Information Analysis Center, Oak Ridge National Laboratory, available online at http://cdiac.esd.ornl.gov/trends/temp/vostok/jouz_tem.htm, and based on the following studies: Jean Jouzel et al., "Vostok Ice Core: A Continuous Isotope Temperature Record Over the Last Climatic Cycle (160,000 Years)," *Nature* 329, no. 6138 (October 1, 1987): 403–8; "Extending the Vostok Ice-Core Record of Palaeoclimate to the Penultimate Glacial Period," *Nature* 364, no. 6436 (July 29, 1993): 407–12; "Climatic Interpretation of the Recently Extended Vostok Ice Records," *Climate Dynamics* 12, no. 8 (June 1996): 513–21; and Jean-Rubert Petit et al., "Climate and Atmospheric History of the

Past 420,000 Years from the Vostok Ice Core, Antarctica," *Nature* 399, no. 6735 (June 3, 1999): 429–36.

16 *"Recent scientific":* National Research Council, *Abrupt Climate Change: Inevitable Surprises* (Washington, D.C.: National Academies Press, 2002). See also Jonathan Adams et al., "Sudden Climate Transitions During the Quaternary," *Progress in Physical Geography* 23 (March 1999): 1–36.

17 *"in periods as short":* Ibid.

17 *"Abrupt climate changes":* Ibid.

18 *white summer ice cap:* Andrew C. Revkin, "Arctic Ice Cap Shrank Sharply This Summer, Experts Say," *New York Times,* September 28, 2005.

18 *2005 study:* Brian Soden et al., "The Radiative Signature of Upper Tropospheric Moistening," *Science* 310, no. 5749 (November 4, 2005): 841–44.

18 *this is precisely:* Petit et al., "Historical Isotropic Temperature."

18 *The warming appears:* Nicholas Caillon et al., "Timing of Atmospheric CO_2 and Antarctic Temperature Changes Across Termination III," *Science* 299, no. 5613 (March 14, 2003); 1728–31.

19 *Figure 2:* Temperature variation from Petit et al., "Historical Isotropic Temperature Record." CO_2 concentrations from Jean-Marc Barnola et al., *Historical CO_2 Record from the Vostok Ice Core,* January 2003, Carbon Dioxide Information Analysis Center, Oak Ridge National Laboratory, available online at http://cdiac.esd.ornl.gov/ftp/trends/co2/vostok.icecore.co2.

19 *a 2-mile-long:* Urs Siegenthaler et al., "Stable Carbon Cycle–Climate Relationship During the Late Pleistocene," *Science* 310, no. 5752 (November 25, 2005): 1313–17.

19 *"That a number of":* Gavin Schmidt, post on RealClimate website, November 24, 2005, www.realclimate.org/index.php?p=221.

20 *Carbon dioxide levels:* Richard Black, "CO_2 'Highest for 650,000 Years,' " BBC News, November 24, 2005, available online at http://news.bbc.co.uk/2/hi/science/nature/4467420.stm.

20 *Eemian interglacial:* James Hansen, "A Slippery Slope," *Climate Change 68,* no. 3 (February 2005): 269–79; and "Is There Still Time."

20 The last time Earth: Hansen, "A Slippery Slope."

21 *1,100 billion tons:* Gregg Marland, Thomas A. Boden, and Robert I. Andres, "Global, Regional, and National CO_2 Emissions," in *Trends: A Compendium of Data on Global Change,* Carbon Dioxide Information Analysis Center, Oak Ridge National Laboratory, U.S. Department of Energy, available online at http://cdiac.esd.ornl.gov/trends/emis/tre_glob.htm.

21 *more than 26 billion tons:* An additional 4 billion tons of carbon dioxide is released annually from land-use changes (mainly burning and decomposition of forest biomass). See Stephen Bernow et al., *The Path to Carbon Dioxide–Free Power* (Washington, D.C.: Tellus Institute and the Center for Energy and Climate Solutions, for the World Wildlife Federation, June 2003).

21 *2 ppm a year:* National Oceanic and Atmospheric Administration ESRL Global Monitoring Division website, http://www.cmdl.noaa.gov/ccgg/ trends/co2_data_mlo.php. Based on Charles David Keeling et al., "Atmospheric Carbon Dioxide Variations at Mauna Loa Observatory, Hawaii," *Tellus* 28 (1976): 538–51; and Kirk W. Thoning, Pieter Tans, and Walter D. Komhyr, "Atmospheric Carbon Dioxide at Mauna Loa Observatory 2: Analysis of the NOAA GMCC Data, 1974–1985," *Journal of Geophysical Research* 94 (1989): 8549–65.

21 *another 0.6°C:* James Hansen et al., "Earth's Energy Imbalance: Confirmation and Implications," *Science* 308, no. 5727 (June 3, 2005): 1431–35, originally published online by *Science* Express, April 28, 2005, www.sciencemag. org/scienceexpress/recent.dtl.

21 *nearly 60 percent:* Hansen, "Is There Still Time."

22 *more than 50 percent:* U.S. Department of Energy, Energy Information Administration, *Annual Energy Outlook 2006,* DOE/EIA 0383 (2006): 71–79, available online at http://www.eia.doe.gov/oiaf/index.html.

22 *2004* Science *magazine article:* Stephen Pacala and Robert Socolow, "Stabilization Wedges: Solving the Climate Problem for the Next 50 Years with Current Technologies," *Science* 305, no. 5686 (August 13, 2004): 968–72.

23 *store that CO$_2$ underground:* Robert Socolow, "Stabilization Wedges: Mitigation Tools for the Next Half-Century" (paper presented at the Global Roundtable on Climate Change, New York, November 14, 2005).

24 *"humanity already possesses":* Pacala and Socolow, "Stabilization Wedges."

26 *climate could be even:* Richard A. Kerr, "Climate Change: Three Degrees of Consensus," *Science* 305, no. 5686 (August 13, 2004): 932–34.

CHAPTER TWO: 2000–2025: REAP THE WHIRLWIND

27 *"I don't see any":* Richard Bradley, "Rain Man," *Plenty,* February/March 2006: 33.

27 *"On our current":* Judith Curry, personal communication, 2006.

28 *On August 28:* The description of Hurricane Katrina in this section is from

Axel Graumann et al., *Hurricane Katrina: A Climatological Perspective,* Technical Report 2005–01, National Climatic Data Center, October 2005, update January 2006; and Richard D. Knabb, Jamie R. Rhome, and Daniel P. Brown, *Tropical Cyclone Report: Hurricane Katrina, 23–30 August 2005.* National Hurricane Center, December 20, 2005.

28 1930s dust bowl: Spencer S. Hsu, "2 Million Displaced by Storms," *Washington Post,* January 13, 2006.

30 *"largely false":* Kerry Emanuel, personal communication, 2006.

30 *Hurricane seasons:* Here I am defining a super-hurricane as any category 4 or 5 hurricane. This is different from the official term *super-typhoon* for typhoons with winds of 150 mph or higher.

31 *extreme events:* World Meteorological Organization, press release, WMO-no. 695, Geneva, July 2, 2003.

31 *"weather is going haywire":* "Reaping the Whirlwind," *Independent* (London), July 3, 2003.

31 *35,000 deaths:* Shaoni Bhattacharya, "European Heatwave Caused 35,000 Deaths," NewScientist.com news service, October 10, 2003, http://www.newscientist.com/article.ns?id=dn4259.

31 *Goddard Institute:* NASA says 2005 just edged out 1998, which it calls "notable" because 1998's temperature was "lifted 0.2°C above the trend line by the strongest El Niño of the past century." See "Global Temperature Trends: 2005 Summation" available online at http://data.giss.nasa.gov/gistemp/2005/.

31 *Mumbai:* Steve Connor, "The Worst Weather Ever? At $200bn, It's Certainly the Costliest," *Independent* (London), December 7, 2005.

31 *"the eight months":* National Oceanic and Atmospheric Administration Satellite and Information Service website, http://lwf.ncdc.noaa.gov/oa/climate/research/2006/jun/hazards.html.

31 *worst wildfire season:* Seth Borenstein, "America's Weather Went Wild in 2005," Knight Ridder newspapers, December 30, 2006; and Wikipedia, s.v. http://en.wikipedia.org/wiki/Northeast_Flooding_of_October_2005.

32 *and more intense—precipitation:* Thomas Karl et al., "Trends in U.S. Climate During the Twentieth Century," *Consequences* 1, no. 1 (Spring 1995).

32 *"precipitation, temperature":* Pavel Ya. Groisman et al., "Contemporary Changes of the Hydrological Cycle over the Contiguous United States: Trends Derived from In Situ Observations," *Journal of Hydrometeorology* 5 (February 2004): 64–85.

32 *precisely what is predicted:* Gabriele Hegerl et al., "Detectability of Anthro-

pogenic Changes in Annual Temperature and Precipitation Extremes," *Journal of Climate,* provisionally accepted.

32 *Washington, D.C.:* Capitalweather.com, June 27, 2006, www.capitalweather. com/2006/06/historic-june-2006-flood-day-three.php.

32 *45 percent:* Sarah Goudarzi, "Scorching U.S.: First Half of 2006 Sets Heat Record," livescience.com, July 14, 2006, www.livescience.com/environment/ 060714_record_heat.html.

32 *Climate Extremes Index:* U.S. Climate Extremes Index, available online at www.ncdc.noaa.gov/oa/climate/research/cei/cei.html.

33 Half or more: *Alaska Native Villages,* General Accounting Office, December 2003, GAO-04-142.

33 *Valdez, Alaska:* Borenstein, "America's Weather Went Wild."

34 *Judith Curry and others:* Judith L. Curry et al., "Mixing Politics and Science in Testing the Hypothesis That Greenhouse Warming Is Causing a Global Increase in Hurricane Intensity," *Bulletin of the American Meteorological Society,* August 2006. See also Judith Curry, "Global Warming and Hurricanes," testimony, U.S. House Committee on Government Reform, *Climate Change Hearing,* July 20, 2006.

34 *"There has been":* Graumann et al., *Hurricane Katrina.*

34 twenty times: Sydney Levitus et al., "Warming of the World Ocean," *Science* 287, no. 5641 (March 24, 2000): 2225–29.

34 *matches the predicted warming:* See also James Hansen et al., "Earth's Energy Imbalance: Confirmation and Implications," *Science* 308, no. 5727 (June 3, 2005): 1431–35.

35 *"A warming signal":* Tim P. Barnett et al., "Penetration of Human-Induced Warming into the World's Oceans," *Science* 309, no. 5732 (July 8, 2005): 284–87.

35 *2006 analysis:* James Hansen et al., "Dangerous Human-Made Interference with Climate: A GISS ModelE Study," 2005, submitted to *Journal of Geophysical Research.* See also Nathan P. Gillett and Peter A. Stott, "Detecting Anthropogenic Influence on Tropical Atlantic Sea Surface Temperatures," *Geophysical Research Abstracts* 8, no. 03698 (2006).

36 *"By trapping heat":* Kerry Emanuel, *Divine Wind: The History and Science of Hurricanes* (New York: Oxford University Press, 2005), p. 23.

36 *2006 report on Katrina:* Graumann et al., *Hurricane Katrina.*

37 *"a methodology":* Carlos D. Hoyos et al., "Deconvolution of the Factors

Contributing to the Increase in Global Hurricane Intensity," *Science* 312, no. 5770 (April 7, 2006): 94–97.

37 *the first, in* Science: Kevin Trenberth, "Uncertainty in Hurricanes and Global Warming," *Science* 308, no. 5729 (June 17, 2005): 1753–54.

38 *at least 74 mph:* For a good discussion of the various terminology used around the world, see Emanuel, *Divine Wind,* 23.

38 *Kerry Emanuel:* Kerry Emanuel, "Increasing Destructiveness of Tropical Cyclones Over the Past 30 Years," *Nature* 436, no. 7051 (August 4, 2005): 686–88.

38 *Georgia Tech:* Peter Webster et al., "Changes in Tropical Cyclone Number, Duration, and Intensity in a Warming Environment," *Science* 309, no. 5742 (September 16, 2005): 1844–46.

39 hypothesis: Dave Wilton, "Theories and Intelligent Design," *A Way with Words: The Weekly Newsletter of Word Origins* 4, no. 12 (June 17, 2005), available online at www.wordorigins.org/AWWW/Vol04/AWWW061705 .html.

39 *National Oceanic and Atmospheric Administration: Climate of 2005: Atlantic Hurricane Season,* National Climatic Data Center, National Oceanic and Atmospheric Administration website, www.ncdc.noaa.gov/oa/climate/ research/2005/hurricanes05.html.

41 *first major critique:* Roger Pielke Jr. et al., "Hurricanes and Global Warming," *Bulletin of the American Meteorological Society,* November 2005. A note indicates "in final form 24 August 2005."

43 *"specifically shows":* Richard Anthes et al., "Hurricanes and Global Warming: Potential Linkage and Consequences," *Bulletin of the American Meteorological Society* May 2006: 623–28.

44 *recent modeling studies:* See, for instance, Hansen et al., "Dangerous Human-Made Interference"; Peter A. Stott et al., "External Control of 20th Century Temperature by Natural and Anthropogenic Forcings," *Science* 290, no. 5499 (December 15, 2000): 2133–37; and Gareth Jones et al., "Sensitivity of Global-Scale Climate Change Attribution Results to Inclusion of Fossil Fuel Black Carbon Aerosol," *Geophysical Research Letters* 32, no. 14 (July 16, 2005).

45 *six major volcanoes:* Alan Robock, "Volcanic Eruptions," in *The Earth System: Physical and Chemical Dimensions of Global Environmental Change,* vol. 1 of *Encyclopedia of Global Environmental Change,* Andrew S.

Goudie and David J. Cuff, eds. (New York: Oxford University Press, 2002), 738–44.

46 *"produced the largest":* Ibid.

46 *Lawrence Livermore National Laboratory:* Peter Gleckler et al., "Volcanoes and Climate: Krakatoa's Signature Persists in the Ocean," *Nature* 439, no. 7077 (February 9, 2006): 675.

47 *"It would appear":* Kerry Emanuel, personal communication, 2006. He pointed out that the "tropical North Atlantic ocean temperature follows the whole northern hemisphere rather closely," which argues against any "regional influence," such as the AMO.

47 *"there is no evidence":* Michael E. Mann and Kerry A. Emanuel, "Atlantic Hurricane Trends Linked to Climate Change," *EOS* 87, no. 24 (June 13, 2006): 233–44. Interestingly, the study notes that the positive or peak phase of the AMO corresponds with the strengthening of the Atlantic's thermohaline circulation, which takes warm and salty water to the coast of western Europe, keeping the continent's climate relatively mild. Some recent evidence suggests that the circulation is weakening, which would mean the AMO is not making a positive contribution to sea-surface temperatures. And that would mean hurricanes are intensifying in spite of—not because of—the AMO.

47 *meteorologist Eric Blake:* Mark Schleifstein, "Katrina Bulks Up to Become a Perfect Storm," *New Orleans Times-Picayune,* August 28, 2005.

47 *"The warmer":* Chris Carroll, "In Hot Water," *National Geographic,* August 2005, 79.

48 *"We think the best": Rush Limbaugh Show,* September 26, 2005, transcript, available online at www.rushlimbaugh.com/home/eibessential/enviro_wackos/max_mayfield_shouts_it_s_not_global_warming.guest.html.

48 *"The increased activity":* Max Mayfield, testimony, *Oversight Hearing on the Lifesaving Role of Accurate Hurricane Prediction,* Senate Committee on Commerce, Science and Transportation Subcommittee on Disaster Prevention and Prediction, 109th Cong., 1st sess. September 20, 2005, available online at www.legislative.noaa.gov/Testimony/mayfieldfinal092005.pdf. See also "NOAA Attributes Recent Increase in Hurricane Activity to Naturally Occurring Multi-Decadal Climate Variability," *NOAA Magazine,* November 29, 2005, available online at www.magazine.noaa.gov/stories/mag184.htm.

48 *a major 2006 study:* Hansen et al., "Dangerous Human-Made Interference."

48 *"a trend in landfalling":* Kerry Emanuel, "Emanuel Replies," *Nature* 438, no. 7071 (December 22, 2005): E13.

49 *"More than half":* Emanuel, personal communication.

49 *"threshold":* Patrick J. Michaels, Paul C. Knappenberger, and Robert E. Davis, "Sea-Surface Temperatures and Tropical Cyclones in the Atlantic Basin," *Geophysical Research Letters* 33, no. 9 (May 10, 2006).

50 *"I don't see":* Kerry Emanuel quoted in Richard Bradley, "Rain Man," *Plenty,* February/March 2006, 33.

50 *Super Typhoon Tip:* National Weather Service Southern Region website www.srh.weather.gov/srh/jetstream/tropics/tc_structure.htm.

50 *Ultimately:* Bill Blakemore, "Category 6 Hurricanes? They've Happened," ABC News, May 21, 2006, available online at http://abcnews.go.com/GMA/print?id=1986862.

51 *Zeta:* "Tropical Depression Zeta Discussion Number 30," National Weather Service Tropical Prediction Center at the National Hurricane Center, Miami, 4 P.M. EST, January 6, 2006. See www.nhc.noaa.gov/archive/2005/dis/al302005.discus.030.shtml.

51 *1°F increase:* Judith Curry and Peter Webster, "Hurricanes & Global Warming" (paper presented at the EESI Symposium *How Changes in the Arctic Are Affecting the Rest of the World,* Washington, D.C., June 15, 2006).

CHAPTER THREE: 2025–2050: PLANETARY PURGATORY

53 *"Obviously":* David Rind, quoted in Elizabeth Kolbert, "The Climate of Man," *The New Yorker,* May 2, 2005.

53 *"We're showing":* Thomas Swetnam, quoted in Tony Davis, "Study: Wildfire Increase Due to Climate Change," *Arizona Daily Star,* July 7, 2006.

53 *Mega-droughts and widespread wildfires:* "NOAA Reports Warmer 2005 for the United States, Near-Record Warmth Globally Hurricanes, Floods, Snow and Wildfires All Notable," *NOAA Magazine,* December 15, 2005, available online at www.noaanews.noaa.gov/stories2005/s2548.htm.

54 *nearly 1°F per decade:* James Hansen et al., "Dangerous Human-Made Interference with Climate: A GISS ModelE Study," 2005, submitted to *Journal of Geophysical Research.*

54 *The oppressive heat:* Shaoni Bhattacharya, "European Heatwave Caused 35,000 Deaths," and "French Heat Toll Tops 11,000," CNN.com, August 29, 2003, www.cnn.com/2003/WORLD/europe/08/29/france.heatdeaths/.

55 *human influence:* Peter Stott et al., "Human Contribution to the European Heat Wave of 2003," *Nature* 432, no. 7017 (December 2, 2004); 610–14.

55 *"These results":* Aiguo Dai et al., "A Global Dataset of Palmer Drought Severity Index for 1870–2002: Relationship with Soil Moisture and Effects of Surface Warming," *Journal of Hydrometeorology* 5 (December 2004): 1117–30.

55 *Every decade:* Millennium Ecosystem Assessment, ed., *Ecosystems and Human Well-Being: Current Status and Trends* (Washington, D.C.: Island Press, 2005), figure 16–8, 449.

55 *"The period since":* Kirk Johnson and Dean Murphy, "Drought Settles In, Lake Shrinks and West's Worries Grow," *New York Times,* May 2, 2004.

56 *Phoenix:* Michael Wilson, "In Phoenix, Even Cactuses Wilt in Clutches of Record Drought," *New York Times,* March 10, 2006. See also "Climate of 2006—June in Historical Perspective," National Climatic Data Center, National Oceanic and Atmospheric Administration website, www.ncdc.noaa .gov/oa/climate/research/2006/jun/jun06.html.

56 *wildfires destroyed:* "Climate of 2005 Wildfire Season Summary" and "Climate of 2006 Wildfire Season Summary," National Climatic Data Center, National Oceanic and Atmospheric Administration website, www.ncdc .noaa.gov/oa/climate/research/2005/fire05.html and www.ncdc.noaa.gov/ oa/climate/research/2006/fire06.html.

56 *A 2005 study:* "Regional Vegetation Die-Off in Response to Global-Change-Type Drought," *Proceedings of the National Academy of Sciences* 102, no. 42 (October 18, 2005): 15144–48. The recent drought had trees dying at rates reaching "90 percent or more" at upper-elevation sites in Colorado and Arizona, whereas the trees that died in the 1950s drought did so mostly at lower elevations.

57 *"We're seeing changes":* Michelle Nijhuis, "Global Warming's Unlikely Harbingers," *High Country News,* July 19, 2004.

57 *thanks in large part:* Kim McGarrity and George Hoberg, "Issue Brief: The Beetle Challenge: An Overview of the Mountain Pine Beetle Epidemic and Its Implications," Department of Forest Resources Management, University of British Columbia, 2005, available online at www.policy.forestry.ubc.ca/ issuebriefs/overview%20of%20the%20epidemic.html.

57 *winter death rate:* "Pine Beetles at 'Epidemic' Levels in Northwest Forests," Associated Press, March 23, 2006, available online at www.signonsandiego. com/news/science/20060323-1411-wst-forestbeetles.html; and The Center

for Health and the Global Environment, *Climate Change Futures* (Harvard Medical School, 2005), available online at www.climatechangefutures.org/pdf/CCF_Report_Final_10.27.pdf.

57 *February 2006 speech:* Senator Lisa Murkowski, "Climate Change: An Alaskan Perspective" (address to Catholic University Law School, Washington, D.C., February 13, 2006), available online at http://murkowski.senate.gov/pdf/Catholic%20U.%20Law%20School.pdf.

57 *Half of the wildfires:* "U.S. Climate Agency Ranks 2005 Near Record for Heat," U.S. Department of State USINFO service, December 16, 2005, available online at http://usinfo.state.gov/gi/Archive/2005/Dec/16-239160.html.

57 *grim reality:* Marvin Eng et al., *Provincial-Level Projection of the Current Mountain Pine Beetle Outbreak: An Overview of the Model (BCMPB v2) and Results of Year 2 of the Project,* for the Mountain Pine Beetle Initiative of the Canadian Forest Service and the British Columbia Forest Service, April 2005, available online at www.for.gov.bc.ca/hre/bcmpb/BCMPB_Main Report_2004.pdf.

57 *"it has become apparent":* McGarrity and Hoberg, "The Beetle Challenge."

57 *"Harvest levels":* Ibid.

58 *"super-interglacial drought":* Jonathan Overpeck, "Warm Climate Abrupt Change—Paleo-Perspectives" (paper presented at the Third Trans-Atlantic Co-operative Research Conference, *Climate, Oceans and Policies—Challenges for the 21st Century,* Washington, D.C., November 1, 2005).

58 *predicted back in 1990:* David Rind et al., "Potential Evapotranspiration and the Likelihood of Future Drought," *Journal of Geophysical Research* 95 (1990): 9,983–10,004 available online at http://pubs.giss.nasa.gov/abstracts/1990/Rind_etal_1.html.

58 *"The development of":* Kirk Johnson et al., "Drought Settles In, Lake Shrinks and West's Worries Grow," *New York Times,* May 2, 2004—through the article makes no mention of global warming.

58 *California's Sierra Nevada:* Robert Service, "As the West Goes Dry," *Science* 303, no. 5661 (February 20, 2004): 1124–27.

59 *2006 study:* Anthony Leroy Westerling, "Warming and Earlier Spring Increases Western U.S. Forest Wildfire Activity," *Science* Express, July 6, 2006, available online at 10.1126/science.1129185.

59 *West is likely:* Steven W. Running, "Is Global Warming Causing More, Larger Wildfires?" (*Science* Express), July 6, 2006, available online at 10.1126/science.1130370.

59 *"the area burned"*: Donald McKenzie et al., "Climatic Change, Wildfire, and Conservation," *Conservation Biology* 18, no. 4 (August 2004): 890–902.

60 *2002 study:* John D. Sterman and Linda Booth Sweeney, "Cloudy Skies: Assessing Public Understanding of Global Warming," *System Dynamics Review* 18, no. 2 (Summer 2002): 207–40.

61 1 ton carbon: The fraction of carbon in carbon dioxide is the ratio of their weights. One ton of carbon, C, equals $44/12 = 11/3 = 3.67$ tons of carbon dioxide, CO_2. The atomic weight of carbon is 12, while the weight of carbon dioxide is 44, because it includes two oxygen atoms that each weigh 16. So, to switch from one to the other, use the formula: 1 ton carbon, C, equals $44/12 = 11/3 = 3.67$ tons carbon dioxide, CO_2.

62 *In 2005, the U.S. Department:* U.S. Energy Information Administration, *Annual Energy Outlook 2006*, DOE/EIA 0383 (2006), available online at www.ela.doe.gov/oiaf/index.html.

65 *"positive feedbacks"*: The climate system does have negative feedbacks loops, whereby a little warming causes a change that slows down warming. For instance, as sea ice retreats, more ocean area is exposed directly to the atmosphere, which could increase the rate at which the ocean takes up carbon dioxide, thereby slowing the rate at which atmospheric concentrations would otherwise have risen. But as the text indicates, scientific observation and analysis strongly suggests the vicious cycles or positive feedbacks dominate the climate system's response to the kind of greenhouse gas forcings it is now experiencing.

65 *"widespread, extreme climatic"*: Appy Sluijs et al., "Subtropical Arctic Ocean Temperatures During the Palaeocene/Eocene Thermal Maximum," *Nature* 441, no. 7093 (June 1, 2006): 610–13.

65 *Middle Ages:* Martin Scheffer et al., "Positive Feedback Between Global Warming and Atmospheric CO_2 Concentration Inferred from Past Climate Change," *Geophysical Research Letter* 33, no. 10 (May 26, 2006).

66 *third study:* Margaret Torn, "Missing feedbacks, asymmetric uncertainties, and the underestimation of future warming," *Geophysical Research Letters* 33, no. 10 (May 26, 2006).

67 *2005 report:* "Oceanic Acidification Due to Increasing Atmospheric Carbon Dioxide," Royal Society (London), June 2005, 7.

67 *more CO_2 would stay:* Jef Huisman et al., "Reduced Mixing Generates Oscillations and Chaos in the Oceanic Chlorophyll Maximum," *Nature* 439, no. 7074 (January 19, 2006): 322–25.

67 *2002 study of Texas:* Richard A. Gill, "Nonlinear Grassland Responses to Past and Future Atmospheric CO_2," *Nature* 417, no. 6886 (May 16, 2002): 279–82.

68 *enlightened energy policies:* John Pickrell, "Soil May Spoil UK's Climate Efforts," *New Scientist* news service, September 7, 2005, available online at http://www.newscientist.com/channel/earth/dn7964-soil-may-spoil-uks-climate-efforts.html; and Pat H. Bellamy et al., "Carbon Losses from All Soils Across England and Wales 1978–2003," *Nature* 437, no. 7056 (September 8, 2005): 245–48.

68 *"locker of carbon":* Laurence Smith, American Meteorological Society seminar, Washington, D.C., February 20, 2006.

68 *nearly 1,000 billion metric tons:* Sergey A. Zimov et al., "Climate Change: Permafrost and the Global Carbon Budget," *Science* 312, no. 5780 (June 16, 2006): 1612–13.

68 *recent degradation:* M. Turre Jorgenson et al., "Abrupt Increase in Permafrost Degradation in Arctic Alaska," *Geophysical Research Letters* 33 (January 24, 2006).

69 *"a mass of shallow lakes":* Fred Pearce, "Climate Warning as Siberia Melts," *New Scientist,* August 11, 2005.

69 *Some 600* million: Wikipedia, s. v. "methane," http://en.wikipedia.org/wiki/Methane.

69 *20 to 60 percent increase:* Torben Christensen et al., "Thawing Sub-Arctic Permafrost: Effects on Vegetation and Methane Emissions," *Geophysical Research Letters* 31 (February 20, 2004).

69 *"the gas was bubbling":* Ian Sample, "Warming Hits 'Tipping Point,' " *Guardian* (London), August 11, 2005.

70 *If concentrations hit 690:* David M. Lawrence and Andrew G. Slater, "A Projection of Severe Near-Surface Permafrost Degradation During the 21st Century," *Geophysical Research Letters* 32 (December 17, 2005); and David Lawrence (American Meteorological Society seminar, Washington, D.C., February 20, 2006). See also Karen E. Frey and Laurence C. Smith, "Amplified Carbon Release from Vast West Siberian Peatlands by 2100," *Geophysical Research Letters* 32 (May 5, 2005).

71 *"at the higher end":* Simon L. Lewis et al., "Tropical Forests and Atmospheric Carbon Dioxide: Current Conditions and Future Scenarios," chapter 14 in *Avoiding Dangerous Climate Change,* eds. Hans Joachim Schellnhuber et al. (Cambridge: Cambridge University Press, 2006).

71 *a 2003* Nature *article:* Mark A. Cochrane, "Fire Science for Rainforests," *Nature* 421, no. 6926 (February 27, 2003): 913–19.

71 *more than 60 feet deep:* Susan E. Page et al., "The Amount of Carbon Released from Peat and Forest Fires in Indonesia During 1997," *Nature* 420, no. 6911 (November 7, 2002): 61–65.

72 *the Amazon was suffering:* Larry Rohter, "A Record Amazon Drought, and Fear of Wider Ills," *New York Times,* December 11, 2005.

72 *Dr. Dan Nepstad:* Fred Pearce, "Amazon Rainforest 'Could Become a Desert,'" *Independent* (London), July 25, 2006.

72 *Models suggest:* Geoffrey Lean, "Dying Forest: One Year to Save the Amazon," *Independent* (London), July 23, 2006.

72 *feedback loop at work:* Peter M. Cox et al., "Amazonian Forest Dieback Under Climate-Carbon Cycle Projections for the 21st Century. *Theoretical and Applied Climatology* 78, no. 1–3 (June 2004): 137–56. See also Richard Betts et al., "The Role of Ecosystem-Atmosphere Interactions in Simulated Amazonian Precipitation Decrease and Forest Dieback Under Global Climate Warming," *Theoretical and Applied Climatology* 78, no. 1–3 (June 2004): 157–75. Tim Flannery, *The Weather Makers: The History & Future Impact of Climate Change* (New York: Atlantic Monthly Press, 2006).

73 *the United States and the world:* Chris D. Jones et al., "Strong Carbon Cycle Feedbacks in a Climate Model with Interactive CO_2 and Sulfate Aerosols," *Geophysical Research Letters* 30 (May 9, 2003): 1479.

73 *important study:* Chris D. Jones et al., "Impact of Climate-Carbon Cycle Feedbacks on Emissions Scenarios to Achieve Stabilisation," chapter 34 in Schellnhuber et al., *Avoiding Dangerous Climate Change* (Cambridge: Cambridge University Press, 2006). This study modeled tundra as if it were any other kind of soil, whereas in fact, as we have seen, it is quite different, especially in its ability to release large amounts of carbon as methane, a far more potent greenhouse gas than carbon dioxide.

CHAPTER FOUR: 2050–2100: HELL AND HIGH WATER

75 *"We could get":* Bob Corell, personal communication.

75 *"The peak rate":* James Hansen, "Defusing the Global Warming Time Bomb," *Scientific American* 290, no. 3 (February 2004): 68–77.

76 *polar amplification:* According to the December 2004 *Arctic Climate Impact Assessment,* a comprehensive report by the leading scientist of the nations that border the Arctic Circle, including ours, over the past 50 years it is

probable, with a confidence level of 66 to 90 percent, that polar amplification has occurred. See also RealClimate website, www.realclimate.org/
index.php?p=234.

77 *"solar heat absorbed":* International Arctic Science Committee (IASC), *Impacts of a Warming Arctic* (Cambridge: Cambridge University Press, 2004),
15.

77 *more than 25 percent:* Andrew Revkin, "In a Melting Trend, Less Arctic Ice
to Go Around," *New York Times,* September 29, 2005.

77 *"At the present rate":* Jonathan Overpeck et al., "Arctic System on a Trajectory to New, Seasonally Ice-Free State, " *Eos* 86, no. 309 (2005): 312–13.

78 *"The recent sea-ice":* Tore Furevik, "Feedbacks in the Climate System and
Implications for Future Climate Projections" (presented at "Climate,
Oceans, and Policies," the Embassy of Norway's Third Annual Forum
Transatlantic Climate Research Conference, Washington, D.C., November
1, 2005).

78 *Most models suggest:* Ola M. Johannessen et al., "Arctic Climate Change—
Observed and Modeled Temperature and Sea Ice Variability," *Tellus* 56A, no.
4 (2004): 328–41.

78 *"0.3° to 0.4°C":* Jonathan A. Foley, "Tipping Points in the Tundra," *Science*
310, no. 5748 (October 28, 2005): 627–28.

78 *A 2005 study:* F. S. Chapin et al., "Role of Land-Surface Changes in Arctic
Summer Warming," *Science* 310, no. 5748 (October 28, 2005): 657–60.

79 *"If this trend persists":* Dr. Wieslaw Maslowski, "Causes of Changes in Arctic
Sea Ice" (paper presented at the American Meteorological Society ESSS
seminar, Washington, D.C., May 2006).

80 *when the planet warms:* Jonathan M. Gregory and Philippe Huybrechts,
"Ice-Sheet Contributions to Future Sea-Level Change," *Philosophical Trans-
actions of the Royal Society* 364, no. 206: 1709–31. Note that some studies
project a faster rate of growth of the Greenland temperatures compared to
global ones. See, for instance, Petr Chylek and Ulrike Lohmann, "Ratio of
the Greenland to Global Temperature Change: Comparison of Observations and Climate Modeling Results," *Geophysical Research Letters* 32, no. 14
(July 21, 2005).

80 *another vicious cycle:* Jonathan Gregory et al., "Threatened Loss of the
Greenland Ice-Sheet," *Nature* 428, no. 6983 (April 8, 2004): 616.

80 *NASA and MIT:* Jay Zwally et al., "Surface Melt-Induced Acceleration of
Greenland Ice-Sheet Flow," *Science* 297, no. 5579 (July 12, 2002): 218–22.

81 *review article in* Science: Julian A. Dowdeswell, "The Greenland Ice Sheet and Global Sea-Level Rise," *Science* 311, no. 5763 (February 17, 2006): 963–64.

81 *1950 to 1996:* Hong-Gyoo Sohn, Kenneth Jezek, and C. J. van der Veen, "Jakobshavn Glacier, West Greenland: 30 Years of Space-Borne Observations," *Geophysical Research Letters* 25, no. 14 (July 15, 1998).

81 *"in October 2000":* Ian Joughin et al., "Large Fluctuations in Speed on Greenland's Jakobshavn Isbrae Glacier," *Nature* 432, no. 7017 (December 2, 2004): 608–10.

81 *A 2006 study:* Adrian Luckman et al., "Rapid and Synchronous Ice-Dynamic Changes in East Greenland," *Geophysical Research Letters* 33 (February 3, 2006). See also "Glacial Change," *Science News* 168 (December 17, 2005): 387.

82 *14 kilometers per year:* "Glacial Change," *Science News* 168 (December 17, 2005): 387.

82 *"accelerated ice discharge":* Eric Rignot and Pannir Kanagaratnam, "Changes in the Velocity Structure of the Greenland Ice Sheet," *Science* 311, no. 5763 (February 17, 2006): 986–90.

82 *"In the next 10 years":* Eric Rignot quoted in Michael D. Lemonick, "Has the Meltdown Begun?" *Time* (February 27, 2006): 38–39.

83 *NASA's Jay Zwally:* "NASA Survey Confirms Climate Warming Impact on Polar Ice Sheets," NASA press release, March 8, 2006.

83 *"The last IPCC report":* Chris Rapley quoted in Matt Weaver, "PM Issues Blunt Warning on Climate Change," *Guardian* (London), January 30, 2006.

83 eight times *as much:* Chris Rapley, "The Antarctic Ice Sheet and Sea Level Rise," in chapter 3 in *Avoiding Dangerous Climate Change,* eds. Hans Joachim Schellnhuber et al. (Cambridge: Cambridge University Press, 2006).

83 *90 percent of Earth's ice:* "Antarctic Ice Sheet Losing Mass, Says University of Colorado Study," ScienceDaily.com, March 2, 2006.

84 *"in the last decade":* ISMASS Committee, "Recommendations for the Collection and Synthesis of Antarctic Ice Sheet Mass Balance Data," *Global and Planetary Change* 42 (2004): 1–15.

84 *The Antarctic Peninsula:* Hamish Pritchard and David G. Vaughan, "Warmer Summers and Faster Glacier Flow on the Antarctic Peninsula" (poster presentation at the Second ENVISAT summer school, Frascati, Italy, August

2004). See also David G. Vaughan et al., "Recent Rapid Regional Climate Warming on the Antarctic Peninsula," *Climatic Change* 60 (2003): 243–74.

84 *lost an area larger:* "Larsen B Ice Shelf Collapses in Antarctica," National Snow and Ice Data Center, March 18, 2002, available online at http://nsidc .org/iceshelves/larsenb2002/.

84 *One glacier's surface:* Eric Rignot et al., "Accelerated Ice Discharge from the Antarctic Peninsula Following the Collapse of Larsen B Ice Shelf," *Geophysical Research Letters* 31, no. 18 (September 22, 2004); and Ted Scambos et al., "Glacier Acceleration and Thinning After Ice Shelf Collapse in the Larsen B Embayment, Antarctica," *Geophysical Research Letters* 31, no. 18 (September 22, 2004).

84 *"the cumulative loss":* Alison Cook et al., "Retreating Glacier Fronts on the Antarctic Peninsula over the Past Half-Century," *Science* 308, no. 5721 (April 22, 2005): 541–44.

85 *"due to an imbalance":* Andrew Shepherd et al., "Warm Ocean Is Eroding West Antarctic Ice Sheet," *Geophysical Research Letters* 31, no. 23 (December 9, 2004); and Fred Pearce, "Antarctic Glaciers Calving Faster into the Ocean," *New Scientist,* October 18, 2005.

85 *A major 2004 study:* Robert Thomas et al., "Accelerated Sea-Level Rise from West Antarctica," *Science* 306, no. 5694 (October 8, 2004): 255–58.

85 *University of Colorado:* "Antarctic Ice Sheet Losing Mass, Says University of Colorado Study," University of Colorado at Boulder, press release, March 2, 2006.

85 *NASA's Zwally:* NASA press release, March 8, 2006.

86 it is fundamentally: "Sea Level, Ice, and Greenhouses—FAQ" available on-line at http://www.radix.net/~bobg/faqs/sea.level.faq.html.

86 *2004 NASA-led study:* Thomas et al. 2004, "Accelerated Sea-Level Rise."

86 *A 2002 study in* Science: Eric Rignot and Stanley S. Jacobs, "Rapid Bottom Melting Widespread Near Antarctic Ice Sheet Grounding Lines," *Science* 296, no. 5575 (June 14, 2002): 2020–23.

86 *another vicious cycle:* As Rapley put it in a 2006 paper, "A combination of accelerated flow and hydrostatic list might cause a runaway discharge." Rapley, "The Antarctic Ice Sheet."

86 *Pine Island and Thwaites:* Pearce, "Antarctic Glaciers."

87 *"A warming of this":* Peter Barrett, "What 3 Degrees of Global Warming Re-ally Means," *Pacific Ecologist* 11 (Summer 2005/06): 6–8.

88 *A 1991 study:* James G. Titus et al., "Greenhouse Effect and Sea Level Rise: The Cost of Holding Back the Sea," *Coastal Management* 19 (1991): 171–204.

88 *The first 1 meter:* James E. Neumann et al., "Sea-Level Rise and Global Climate Change: A Review of Impacts to U.S. Coasts" (prepared for the Pew Center on Global Climate Change, February 2000).

89 *One 2001 analysis:* James G. Titus and Charlie Richman, "Maps of Lands Vulnerable to Sea Level Rise: Modeled Elevations Along the U.S. Atlantic and Gulf Coasts," *Climate Research* 18 (2001): 205–28.

89 *they don't consider the impact:* Stephen Schneider and Robert S. Chen, "Carbon Dioxide Warming and Coastline Flooding: Physical Factors and Climatic Impact," *Annual Review of Energy* 5 (November 1980): 107–40.

89 *a world where sea levels:* Peter Whoriskey, "Post-Katrina Rebuilders Hug Ground, Trust Levees," *Washington Post*, February 26, 2006.

90 *2005 study:* Robert J. Nicholls, Richard S. J. Tol, and Nassos Vafeidis, "Global Estimates of the Impact of a Collapse of the West Antarctic Ice Sheet," January 6, 2004, available online at www.uni-hamburg.de/Wiss/FB/15/Sustain ability/annex6.pdf.

90 *A 1980 paper:* Schneider and Chen, "Carbon Dioxide Warming."

CHAPTER FIVE: HOW CLIMATE RHETORIC
TRUMPS CLIMATE REALITY

99 *"The scientific debate":* Frank Luntz, "Straight Talk" memo (Luntz Research Companies, Washington, D.C., 2002), 131–46, available online at http://www.politicalstrategy.org/archives/001330.php.

99 *"Global warming":* David Brooks, "Running Out of Steam," *New York Times*, December 8, 2005.

100 *Kyoto Protocol:* Different countries have different targets. The 5% figure is the average. See Wikipedia, s. v. "Kyoto Protocol," http://en.wikipedia.org/wiki/Kyoto_Protocol.

100 *"It is clear":* Tony Blair, speech on sustainable development, February 2003, available online at www.number-10.gov.uk/output/Page3073.asp.

101 *"And by long-term":* Tony Blair, speech on climate change, London, September 14, 2004, available online at www.number10.gov.uk/output/page6333.asp.

101 *"squandering the chance":* "UK Must Lead on Climate Change," BBC News, September 13, 2004, available online at http://news.bbc.co.uk/1/hi/uk_

politics/3651052.stm. Also see "UK: PM Gives Dire Warning on Climate," BBC News, September 15, 2004, available online at http://news.bbc.co .uk/1/hi/uk_politics/3656812.stm.

101 *major Senate bill:* McCain-Lieberman climate bill roll-call vote is available online at www.senate.gov/legislative/LIS/roll_call_lists/roll_call_vote_cfm .cfm?congress=109&session=1&vote=00148.

102 *"You need 60":* "US Senate Likely to Reject Future UN Climate Deal—Interview," EurActiv.com, February 15, 2006.

102 *"It is clear":* Tony Blair, Foreword, in *Avoiding Dangerous Climate Change,* eds. Hans Joachim Schellnhuber et al. (Cambridge: Cambridge University Press, 2004).

102 *"Though he didn't":* Fred Barnes, *Rebel-in-Chief: Inside the Bold and Controversial Presidency of George W. Bush* (New York: Random House, 2006).

103 *"a compelling presentation":* Senator James M. Inhofe, "Climate Change Update," Senate floor statement, 109th Cong., 1st sess., January 4, 2005, available online at http://inhofe.senate.gov/pressreleases/climateupdate.htm. See also http://epw.senate.gov/hearing_statements.cfm?id=246814.

103 *"Scientists agree":* John Tierney, "And on the Eighth Day, God Went Green," *New York Times,* February 11, 2006.

105 *"Un-Journalism":* Jude Wanniski, "Un-Journalism at the New Yorker," May 9, 2005, available online at www.wanniski.com/showarticle.asp?articleid=4350. See also wanniski.com/PrintPage.asp?TextID=3550.

105 *"There is no relationship":* Charles Krauthammer, "Where to Point the Fingers," *Washington Post,* September 8, 2005.

106 *the recent scientific evidence:* "Will Railed About Global Warming-Hurricane Link Claim; Ignored Actual Scientific Data on Hurricane Intensity," *Media Matters for America,* September 26, 2005, available online at http://media matters.org/items/200509260004.

106 *"phony theory":* Charles Krauthammer, "Phony Theory, False Conflict," *Washington Post,* October 18, 2005.

106 *"But it is":* George F. Will, "Grand Old Spenders," *Washington Post,* November 17, 2005.

107 *"Crichton's subject":* George F. Will, "Global Warming? Hot Air," *Washington Post,* December 23, 2004.

108 *"Of all the talents":* Winston Churchill, "The Scaffolding of Rhetoric," unpublished essay, 1897.

108 *"Aptness of language"*: Aristotle, *Rhetoric*, cited in Brian Vickers, *Classical Rhetoric in English Poetry* (Carbondale: Southern Illinois University Press, 1970), 94.

108 *"constitute basic schemes"*: Raymond W. Gibbs, Jr., *The Poetics of Mind* (Cambridge: Cambridge University Press, 1994), 1.

109 *"All the speeches"*: Churchill, "Scaffolding."

109 *"There's a simple"*: Frank Luntz, interview on PBS's *Frontline*, November 9, 2004, available online at www.pbs.org/wgbh/pages/frontline/shows/persuaders/themes/citizen.html.

110 *"Scientists have"*: Mark Bowen, *Thin Ice Unlocking the Secrets of Climate in the World's Highest Mountains* (New York: Henry Holt, 2005), 21.

111 *Words alone*: Royal Society website, www.royalsoc.ac.uk/page.asp?id=1020.

111 *"Scientists who do"*: Jared Diamond, "Kinship with the Stars," *Discover* 18 (May 1997): 44–49.

111 *"For a scientist"*: Judith Curry et al., "Mixing Politics and Science in Testing the Hypothesis That Greenhouse Warming Is Causing a Global Increase in Hurricane Intensity," *Bulletin of the American Meteorological Society*, August 2006.

112 *does not melt*: "Governor Schwarzenegger Announces Landmark GHG Reduction Goals," June 2005, available online at www.climateregistry.org/Default.aspx?TabID=3423&refreshed=true.

112 *"one who takes"*: Answers.com. s.v. "contrarian," www.answers.com/topic/contrarian.

113 *"If you just"*: Michael Crichton, in Michael Crowley, "Michael Crichton's Scariest Creation: Jurassic President," *New Republic*, March 20, 2006.

113 *2002 memo*: Luntz, "Straight Talk."

114 *emphasis in original throughout*: I use italics here and throughout to signify emphasis for Luntz, but he often uses multiple emphases combining italics with boldface, and sometimes combining both of those with underlining.

114 *the phrase "climate change"*: Andrew C. Revkin, "Call for Openness at NASA Adds to Reports of Pressure," *New York Times*, February 16, 2006.

115 *For Luntz and a large*: In 2006, Luntz was asked by the BBC about the memo and replied, "It's now 2006. Now I think most people would conclude that there is global warming taking place, and that the behavior of humans are affecting the climate." But that was true in 2002. And in any case his cynical lines—"The scientific debate is closing (against us) but not yet closed. There is still a window of opportunity to challenge the science."—imply

that he knew he was on the losing side of the issue scientifically but believed the issue could still be won rhetorically. See "Luntz Converts on Global Warming, Distances Himself from Bush," available online at http://think progress.org/2006/06/27/luntz-gw/.

116 *"Doubt is":* Tobacco memo available online at www.prevention.ch/doubt-is-our-product.pdf.

116 *"how much warming":* Will, "Global Warming?"

117 *"We must not rush":* Luntz, "Straight Talk."

117 *"Science tells us":* Paula Dobriansky, remarks to "The Convention After 10 Years: Accomplishments and Future Challenges," *Tenth Session of the Conference of the Parties (COP) to the U.N. Framework Convention on Climate Change,* Buenos Aires, Argentina, December 15, 2004, available online at www.uspolicy.be/Article.asp?ID=C4A8C67B-E36F-45EA-B557-EEF 7F9A6EB4A.

118 *stunning conclusion:* Government Accountability Office, *Climate Change Assessment: Administration Did Not Meet Reporting Deadline,* report to Senator John McCain and Senator John Kerry, April 14, 2005. See also Andrew C. Revkin, "Climate Research Faulted Over Missing Components," *New York Times,* April 22, 2005.

118 *White House had secretly:* "Group Sues to Enforce Sound Science Law," Competitive Enterprise Institute, press release, August 6, 2003. See also Ross Gelbspan, *Boiling Point: How Politicians, Big Oil and Coal, Journalists and Activists Are Fueling the Climate Crisis—and What We Can Do to Avoid Disaster* (New York: Basic Books, 2004), 56–58.

118 *White House heavily:* "White House Guts Global Warming Study," CBS News, June 19, 2003, available online at www.cbsnews.com/stories/ 2003/07/24/politics/main564873.shthml; and Andrew C. Revkin and Katharine Q. Seelye, "Report by the E.P.A. Leaves Out Data on Climate Change," *New York Times,* June 19, 2003.

119 *His documents showed:* Climate Change Research Distorted and Suppressed," Union of Concerned Scientists website, at www.ucsusa.org/ scientific_integrity/interference/climate-change.html. This page contains excerpts from *Scientific Integrity in Policymaking,* Union of Concerned Scientists, 2004.

119 *More recently:* Andrew C. Revkin, "Climate Expert Says NASA Tried to Silence Him," *New York Times,* January 29, 2006. See also John B. Judis, "The Government's Junk Science: NOAA's Flood," *National Review,* February 20,

2006; and "Rewriting the Science," *60 Minutes,* CBS News, July 30, 2006, available online at www.cbsnews.com/stories/2006/03/17/60minutes/main 1415985.shtml.

120 *"Scientists who don't":* Judis, "The Government's Junk Science."

120 *"I do believe":* Conrad Lautenbacher, quoted in ibid.

121 *"People have hunches":* Conrad Lautenbacher, quoted in Bill Lambrecht, "Missourians Should Heed Storm Lesson, Experts Say," *St. Louis Post-Dispatch,* August 31, 2005.

121 *"a few recent":* "Former NOAA Lab Director: 'Climate Scientists Within NOAA Have Been Prevented from Speaking Freely,' " ClimateScienceWatch post, March 10, 2006, available online at www.climatesciencewatch.org/index.php/csw/details/mahlman-lautenbacher/.

121 *"Contrary to Dr. Lautenbacher's":* Jerry Mahlman, personal communication, 2006.

122 *"With all of the":* Chris Mooney, "Earth Last," *American Prospect,* May 4, 2004.

122 *A 1977 report:* Walter Sullivan, "Scientists Fear Heavy Use of Coal May Bring Adverse Shift in Climate," *New York Times,* July 25, 1977. The rest of the history is available online at www.aip.org/history/climate/Govt.htm.

123 *"In the 1970s":* Michael Crichton, *State of Fear* (New York: HarperCollins, 2004), 315.

123 *George Will picked:* Will, "Global Warming?"

124 *"The Global Cooling Myth":* RealClimate website, http://www.realclimate.org/index.php?p=94. One quote that Will ascribes to the prestigious peer-reviewed journal *Science* actually came from the non-peer-reviewed magazine *Science News.* For a detailed debunking of the notion that scientists were predicting an imminent ice age in the 1970s, see www.wmconnolley.org.uk/sci/iceage/.

124 *A spring 2003 workshop:* Fred Pearce, "Global Warming's Sooty Smoke-screen Revealed," *New Scientist,* June 4, 2003.

124 *A 2005 study:* Thomas L. Delworth et al., "The Impact of Aerosols on Simu-lated Ocean Temperature and Heat Content in the 20th Century," *Geophys-ical Research Letters* 32 (December 21, 2005).

125 *"Truthiness":* "The Colbert Report," *60 Minutes,* CBS News, April 30, 2006, available online at www.cbsnews.com/stories/2006/04/27/60minutes/main 1553506.shtml.

125 *"STATE OF FEAR":* "State of Fear," Marich Communications, press release,

December 7, 2004, available online at www.michaelcrichton.com/press/index/html.

125 *geneticists were executed:* Wikipedia, s.v. "Lysenkoism," http://en.wikipedia.org/wiki/Lysenkoism.

126 *"In light of":* Science 304, no. 5669 (April 16, 2004): 400–402.

126 *"Hansen overestimated":* James Hansen, "The Global Warming Debate," NASA website, January 1999, available online at www.giss.nasa.gov/edu/gwdebate.

126 *Michaels is:* Cato Institute website, www.cato.org/people/michaels.html.

128 *The environmentalists did:* "Contextomy Tsunami" at http://www.fallacyfiles.org/archive012005.html; and Bill McKibben, "Stranger Than Fiction," *Mother Jones,* May/June 2005.

128 *"The Tsunami Exploiters":* James Glassman, "The Tsunami Exploiters," Tech Central Station, January 14, 2005, available online at www.techcentralstation.com/011405C.html.

129 *"The Death of Environmentalism":* Michael Shellenberger and Ted Nordhaus, "The Death of Environmentalism," September 2004, available online at www.thebreakthrough.org/images/Death_of_Environmentalism.pdf.

CHAPTER SIX: THE TECHNOLOGY TRAP AND THE AMERICAN WAY OF LIFE

133 *"There is no doubt":* Tony Blair, "The Prime Minister's Speech to the Business and Environment Programme," September 14, 2004, available online at www.g8.gov.uk/servlet/Front?pagename=OpenMarket/Xcelerate/ShowPage&c=Page&cid=1078995903270&aid=1097485779120.

133 *"It's important":* Mark Hertsgaard, "While Washington Slept," *Vanity Fair,* April 17, 2006.

135 *"What will never fly":* Shankar Vedantam, "Senate Impasse Stops 'Clear Skies' Measure," *Washington Post,* March 10, 2005.

135 *"to ensure that":* Fiona Harvey, "U.S. Is Accused of Undermining Kyoto Principles on Emissions," *Financial Times,* December 17, 2004.

138 *"voluntary programs":* Conservatives often use the phrase "voluntary programs" to mean efforts to get industry to make voluntary pledges to reduce emissions.

139 *"The United States is":* Spencer Abraham, quoted in "U.S. Energy Secretary Says New Technologies Needed to Achieve Global Climate Goals," U.S.

Newswire September 17, 2003, available online at http://releases.usnews wire.com/GetRelease.asp?id=20881.

140 *"With a new":* George W. Bush, State of the Union address, January 28, 2003.

140 *A hydrogen car:* Joseph Romm, *The Hype About Hydrogen: Fact and Fiction in the Race to Save the Climate* (Washington, D.C.: Island Press, 2005).

140 *A 2005 Luntz:* Frank Luntz, "An Energy Policy for the 21st Century," *A New American Lexicon,* March 2005, available online at www.politicalstrategy. org/archives/001207.php#1207.

140 *"What's most striking":* "President Discusses Energy at National Small Business Conference," White House press release, April 27, 2005; and John Carey, "Bush Is Blowing Smoke on Energy," *Business Week,* April 28, 2005.

141 *"Sometimes, decisions":* "Bush Blames 'Mixed Signals' for Energy Lab Layoffs," *USA Today,* February 21, 2006.

143 *"force for good":* "Put a Tiger in Your Think Tank," *Mother Jones,* May/June 2005, available online at www.motherjones.com/news/featurex/2005/05/ exxon_chart.html.

143 *"to depict global":* "Industrial Group Plans to Battle Climate Treaty," *New York Times,* April 26, 1998.

146 *Over the next few years:* "High Temperature Superconductors," available online at www.eapen.com/jacob/superconductors/chapter5.html.

146 *"Typically it has":* Global Business Environment, *Energy Needs, Choices and Possibilities: Scenarios to 2050* (London: Shell International, 2001), 22.

146 *We barely have:* Romm, *Hype.*

147 *Research on nickel:* Battery University.com, www.batteryuniversity.com/ partone-4.htm.

149 "Are we going": Stephen Johnson, quoted in Juliet Eilperin, "Ex-EPA Chiefs Agree on Greenhouse Gas Lid," *Washington Post,* January 19, 2006.

150 *$10 to $20 billion per year:* Daniel N. Kammen and Gregory F. Nemet, "Reversing the Incredible Shrinking Energy R&D Budget," *Issues in Science and Technology* (Fall 2005): 84–88. Missile Defense funding numbers from Missile Defense Agency FY07 budget estimate, available online at www.cdi .org/pdfs/Final%20Budget%20Overview%20FY%202007%20MDA.pdf.

152 *"the intential large-scale manipulation":* David Keith, "Geoengineering Climate," *Elements of Change,* S. J. Hassol and J. Katzenberger, eds. (Aspen, Colo.: Aspen Global Change Institute, 1998), 83–88.

152 *"The 'geo-engineering' "*: John Holdren, "The Energy Innovation Impera-
tive," *Innovations* 1, no. 2 (Spring 2006): 3–23.

CHAPTER SEVEN: THE ELECTRIFYING SOLUTION

154 *"This analysis suggests"*: National Academy of Sciences, *Policy Implications
of Greenhouse Warming: Mitigation, Adaptation, and the Science Base*
(Washington, D.C.: National Academies Press, 1991).

155 *The coal plants that will:* David G. Hawkins, testimony, U.S. House Committee on Energy and Commerce, Subcommittee on Energy and Air Quality, *Hearing on Future Options for Generation of Electricity from Coal,* 108th
Cong., 1st sess., June 24, 2003, available online at www.nrdc.org/global
Warming/tdh0603.asp. The new plants amount to some 1,400 GW, which
includes 400 GW of plants to replace existing ones that have reached the
end of their lifetime.

156 *The total extra costs:* U.S. Department of Energy, Office of Fossil Energy,
Carbon Sequestration R&D Overview, available online at www.fe.doe.gov/
programs/sequestration/overview.html. This is the cost for large-scale sequestration in places like deep underground aquifers. Small-scale sequestration for enhanced oil and gas recovery is far less expensive.

156 *"Vendors currently"*: *Coal-Related Greenhouse Gas Management Issues,* National Coal Council, Washington, D.C., May 2003.

157 *FutureGen project:* U.S. Department of Energy, Office of Fossil Energy, *FutureGen Fact Sheet,* available online at www.fossil.energy.gov/programs/
powersystems/futuregen/.

158 *"Less dense"*: National Research Council, *Novel Approaches to Carbon Management,* Workshop Report (Washington, D.C.: National Academies Press,
2003), 3.

158 *Pacific Northwest National:* James Dooley and Marshall Wise, "Why Injecting CO_2 into Various Geologic Formations Is Not the Same as Climate
Change Mitigation: The Issue of Leakage," Joint Global Change Research
Institute (Battelle Pacific Northwest National Laboratory), 2002. See also
David Hawkins, "Passing Gas: Policy Implications of Leakage from Geologic Carbon Storage Sites," Natural Resources Defense Council, Washington, D.C., 2002.

159 *Analysis suggests:* See, for instance, Keith and Farrel; and Timothy Johnson
and David Keith, "Fossil Electricity and Carbon Dioxide Sequestration,"
Energy Policy 32, no. 4 (March 2004): 367–82. See also Howard Herzog,

"The Economics of CO_2 Separation and Capture," *Technology* 7, supp. 1 (2000): 13–23.

159 *Energy efficiency remains:* See, for instance, Arthur H. Rosenfeld, "Sustainable Development—Reducing Energy Intensity by 2 percent Per Year" (PowerPoint presentation at the International Seminar on Planetary Emergencies, Erice, Italy, August 19, 2003).

160 *This astonishing achievement:* The chart is derived from "Consumption, Physical Units, 1960–2002" for electricity consumption (kWh) and from "Appendix C: Resident Population" of U.S. Department of Energy, "Data Sources and Technical Notes," available online at http://www.eia.doe.gov/emeu/states/_seds.html.

163 *California utilities:* Cynthia Rogers et al., "Funding and Savings for Energy Efficiency Programs for Program Years 2000 through 2004," paper, California Energy Commission Staff, July 2005.

163 *2006 report:* Western Governors' Association, "Energy Efficiency Task Force Report," January 2006, available online at www.westgov.org/wga/initiatives/cdeac/Energy%20Efficiency-full.pdf.

164 And it is avoiding: Personal communications with Art Rosenfeld. See also Audrey Chang, "California's Sustainable Energy Policies Provide a Model for the Nation," Natural Resources Defense Council, Washington, D.C., May 2005. Available online at http://www.e2.org/ext/doc/CASustEnergy Policies.pdf.

165 *increased demand for gas:* Energy Information Administration, *Annual Energy Outlook 2003,* DOE/EIA-0383 (2006): 67.

166 *Steam accounts:* "BestPractices Steam," Alliance to Save Energy website, www.ase.org/section/program/bpsteam. See also Joseph Romm, *Cool Companies: How the Best Businesses Boost Profits and Productivity by Cutting Greenhouse Gas Emissions* (Washington, D.C.: Island Press, 1999).

166 *2 million new jobs:* Joseph Romm and Charles Curtis, "Mideast Oil Forever?" *Atlantic Monthly,* April 1996. According to the study, this is "a relatively large impact considering that the investments driving it were assumed to be made for purposes other than increasing employment."

167 *about 88,000 megawatts:* "The Market and Technical Potential for Combined Heat and Power in the Commercial/Institutional Sector," prepared by Onsite Sycom Energy Corp. for the U.S. Department of Energy, Washington, D.C., January 2000, available online at www.eere.energy.gov/de/pdfs/chp_comm_market_potential.pdf.

168 *July 2000 report:* R. Brent Alderfer et al., *Making Connections: Case Studies of Interconnection Barriers and Their Impact on Distributed Power Projects* (Golden, Colo., National Renewable Energy Laboratory, July 2000).

169 *9,000 megawatts:* John Douglas, "Putting Wind on the Grid," *EPRI Journal* (Spring 2006): 6–15.

170 *"2–5 cents/kilowatt-hour":* *Renewables 2005 Global Status Report: Notes and Reference Companion Document,* Renewable Energy Policy Network for the 21st Century (REN21), October 20, 2005, available online at www.ren21 .net/globalstatusreport/RE2005_Notes_References.pdf.

171 *The E.U. has set:* Ibid, pp. 19–24.

171 *Department of Energy study:* U.S. Department of Energy, Energy Information Administration, *Analysis of Strategies for Reducing Multiple Emissions from Electric Power Plants: Sulfur Dioxide, Nitrogen Oxides, Carbon Dioxide, and Mercury and a Renewable Portfolio Standard,* SR/OIAF/2001–03, July 2001. Electricity prices in 2020 under a 20 percent RPS would be about 4 percent higher than the EIA projects they would be in a business-as-usual scenario, but 2 percent lower than they are today.

171 *"The Path to":* Alison Bailie et al., *The Path to Carbon-Dioxide-Free Power: Switching to Clean Energy in the Utility Sector* (Washington, D.C.: Tellus Institute and Center for Energy and Climate Solutions, Report for the World Wildlife Federation 2003), available online at http://assets.panda .org/downloads/powerswitchfinalusa.pdf.

172 *The net savings:* Net savings included the costs for more energy-efficient equipment and additional cogeneration (plus transfers of revenue from the CO_2 cap and trade program back to the consumers).

174 *"It's the only":* Darren Samuelsohn, "McCain Says White House Run Would Not Change Commitment to Emission Curbs," *E&E News,* March 15, 2006.

174 *Nuclear energy is:* Notwithstanding the fact that uranium enrichment in this country makes use of a highly electricity-intensive process that is almost exclusively powered by coal plants.

175 *2003 study by MIT:* John Deutch et al., *The Future of Nuclear Power* (Cambridge: Massachusetts Institute of Technology, 2003).

175 *"The abiding lesson":* Matthew L. Wald, "Interest in Building Reactors, but Industry Is Still Cautious," *New York Times,* May 2, 2005.

176 *California, however: Energy Source, CA Total Electric Power Industry Net Generation,* U.S. Department of Energy, Energy Information Administra-

tion data table, available online at www.eia.doe.gov/cneaf/solar.renewables/page/state_profile/rsp_ca_table3.html.

CHAPTER EIGHT: PEAK OIL, ENERGY SECURITY, AND THE CAR OF THE FUTURE

177 *"We have a serious"*: George W. Bush, State of the Union address, January 31, 2006.

177 *"In the absence"*: Senator Richard Lugar, "Energy: The Albatross of National Security," submitted to *Conservative Environmental Policy—Quarterly*, spring 2006, available online at http://lugar.senate.gov/energy/press/articles/060301cepquarterly.html.

178 *"Our nation"*: George W. Bush, quoted in "Bush Pushes Alternative Energy Proposals," Associated Press, February 20, 2006, available online at www.msnbc.msn.com/id/11465801/.

179 *And the transportation sector*: U.S. Department of Energy, Energy Information Administration, *Annual Energy Outlook 2003*, Table A19, DOE/EIA= 0383 (2003). Available online at www.eia.doe.gov/oiaf/archive/aeo03/index.html.

179 *"In the absence"*: "Biofuels for Transport: An International Perspective," International Energy Agency, press release, May 11, 2004, available online at www.iea.org/Textbase/press/pressdetail.asp?PRESS_REL_ID=127.

180 *more oil than the entire world*: Romm and Curtis, "Mideast Oil Forever?" *Atlantic Monthly*, April 1996.

180 *when prices spiked*: Energy Information Administration, *International Petroleum Monthly*, March 2006.

180 *"There is nothing"*: Kenneth Deffeyes, *Hubbert's Peak: The Impending Oil Shortage* (Princeton, N.J.: Princeton University Press, 2001), 158. For the opposing view, see Leonardo Maugeri, "Oil: Never Cry Wolf—Why the Petroleum Page Is Far from Over," *Science* 304, no. 5674 (May 21, 2004): 1114–15.

180 *"A scarcity of oil"*: Global Business Environment, *Energy Needs, Choices and Possibilities: Scenarios to 2050* (London: Shell International, 2001), 18.

181 *200 to 400 billion*: International Energy Agency, *Resources to Reserves*, Paris, September 2005; and David Adam, "Global Warming Sparks a Scramble for Black Gold Under Retreating Ice," *Guardian* (London), April 18, 2006.

181 *heavy oil in Venezuela*: Manik Talwani, "The Orinoco Heavy Oil Belt in Venezuela (or Heavy Oil to the Rescue?)," Rice University, Houston, Texas, September 2002, available online at http://cohesion.rice.edu/natural

sciences/earthscience/research.cfm?doc_id=2819; and *Alberta's Oil Sands 2004,* Government of Alberta, Ministry of Energy, available online at www .energy.gov.ab.ca/docs/oilsands/pdfs/PUB_osgenbrf.pdf.

182 *Canada's increasing:* The U. S. Energy Information Administration projects a sharp decline in net imports of Canadian natural gas by 2020. EIA, U.S. Department of Energy, Energy Information Administration *Annual Energy Outlook 2006,* DOE/EIA (2006). See also U.S. DOE, EIA, *Annual Energy Outlook 2004* DOE/EIA (2004), p. 50.

182 *Colorado and Utah:* James R. Udall and Steven B. Andrews, "The Illusive Bonanza: Oil Shale in Colorado," *Energy Bulletin* (October 3, 2005).

183 *"CO$_2$ flooding":* Robert Hirsch et al., *Peaking of World Oil Production,* Science Applications International Corp., February 2005.

184 *21 million barrels:* Northern Plains Resource Council, *Montana's Energy Future,* 2006; available online at http://www.worc.org/pdfs/Synfuel_Briefing_ Paper.pdf; and David Garman, "Unconventional Liquid Fuels" (PowerPoint presentation to Defense Science Board, U.S. Department of Energy, Washington, D.C., June 2006).

184 *Worse, the total:* Adam Brandt and Alexander Farrell, "Scraping the Bottom of the Barrel: Greenhouse Gas Emission Consequences of a Transition to Low-Quality and Synthetic Petroleum Resources," submitted to *Climatic Change.*

185 *2 billion metric tons:* Ibid.

185 *"Forget hydrogen":* James Woolsey, remarks at Plug-in America press conference, National Press Club, Washington, D.C., January 2006, available online at www.connectlive.com/events/austinenergy/.

185 *"$1.2 billion":* George W. Bush, State of the Union address, 2003, available online at www.whitehouse.gov/news/releases/2003/01/20030128-19 .html.

185 *"It is highly likely":* National Research Council and National Academy of Engineering, *The Hydrogen Economy: Opportunities, Costs, Barriers, and R&D Needs* (Washington, D.C.: National Academies Press, 2004).

186 *January 2004 study: Well-to-Wheels Analysis of Future Automotive Fuels and Powertrains in the European Context,* European Commission Center for Joint Research, EUCAR, and Concawe, Brussels, January 2004.

186 *save four times:* For a longer discussion of hydrogen cars and plug-in hybrids, see Joseph Romm, *The Hype About Hydrogen: Fact and Fiction in the Race to Save the Climate* (Washington, D.C.: Island Press, 2005).

186 *a pollution-free:* Internal combustion engine cars can also be modified to run on hydrogen, although they are considerably less efficient than fuel-cell vehicles and thus have much shorter range and even higher annual fuel bills.

186 *currently cost about $2,000:* U.S. Department of Energy, *Basic Research Needs for the Hydrogen Economy* (Washington, D.C.: Office of Science, 2003).

187 *"a new material":* American Physical Society, "The Hydrogen Initiative," March 2004.

187 *more than $500 billion:* Marianne Mintz et al., "Cost of Some Hydrogen Fuel Infrastructure Options" (Argonne National Laboratory, presentation to the Transportation Research Board, Washington, D.C., January 16, 2002).

187 *"Fuel-cell cars":* Matt Wald, "Questions About a Hydrogen Economy," *Scientific American* 290, no. 5 (May 2004): 66–73.

187 *"If I told you":* Bill Reimert, quoted in Jamie Butters et al., "Fuel-Economy Technologies," *Detroit Free Press,* January 10, 2005.

187 *2004 MIT study:* Nancy Stauffer, *New Vehicle Technologies: How Soon Can They Make a Difference?* (Cambridge: MIT Laboratory for Energy and the Environment, 2005). Available online at http://esd.mit.edu/esd_reports/summer2005/new_vehicle_technologies.html.

187 *5-year budget:* Daniel Whitten, "Barton Rails at Budget Request for Shorting EPACT," *Inside Energy,* March 13, 2006.

188 *Samuel Bodman announced:* Geoff Brumfiel, "Energy Secretary Ditches Science Advisers," *Nature* 440, no. 7085 (April 6, 2006): 725.

188 *"My message":* Lugar, "Energy: The Albatross."

189 *European countries: Transportation Energy Data Book,* edition 22 (Oak Ridge, Tenn.: Oak Ridge National Laboratory, 2002), 5-2, 5-3.

189 *nowhere near 60 mpg:* Feng An and Amanda Sauer, *Comparison of Passenger Vehicle Fuel Economy and Greenhouse Gas Emissions Standards Around the World* (Arlington, Va.: Pew Center on Global Climate Change, December 2004). As of 2002, the average fuel economy of European Union vehicles was 37 mpg, and some of that fuel-economy improvement was achieved not just with high fuel prices but with strong tax incentives to promote diesel vehicles, which are typically more fuel-efficient.

189 *In a 2002 report:* National Research Council, *Effectiveness and Impact of Corporate Average Fuel Economy (CAFE) Standards* (Washington, D.C.: National Academies Press, 2002).

189 *Studies by the national:* Interlaboratory Working Group, *Scenarios of U.S. Carbon Reductions,* Lawrence Berkeley National Laboratory and Oak Ridge National Laboratory, prepared for the Office of Energy Efficiency and Renewable Energy, U.S. Department of Energy, September 1997, pp. 5.44–5.48; David Greene and Andreas Schafer, *Reducing Greenhouse Gas Emissions from U.S. Transportation* (Arlington, Va.: Pew Center on Global Climate Change, May 2003), 13–18; and Malcolm Weiss et al., "On the Road in 2020: A Life-Cycle Analysis of New Automobile Technologies" (Cambridge: MIT, October 2000), tables 5.3 and 5.4.

   An Oak Ridge National Laboratory study found that "based on a comparison of fatality data for SUVs to other vehicles, the registered-vehicle-fatality rate (defined as number of fatalities per number of registered vehicles) for SUVs is higher than the registered-vehicle-fatality rate for other vehicles." Stacy Davis and Lorena Truett, *An Analysis of the Impact of Sport Utility Vehicles in the United States,* ORNL (Oak Ridge, Tenn.: Oak Ridge National Laboratory, 2000), 24.

190 *"Policies aimed":* Robert Noland, "Fuel Economy and Traffic Fatalities," *Energy Policy* 33 (2005): 2183–90.

190 *Toyota Prius hybrid:* "Toyota Prius: AEI Best Engineered Vehicle 2004," *Automotive Engineering International,* March 2004: 58–68.

190 *Europeans have still:* An and Sauer, *Passenger Vehicle Fuel Economy.* I have used their normalization so that European mpg can be directly compared with mpg calculated under CAFE.

190 Consumer Reports *found:* "Fuel Economy: Why You're Not Getting the MPG You Expect," *Consumer Reports,* October 2005.

191 *Center for American Progress:* Bracken Hendricks et al., "Strengthening America's Auto Industry," Center for American Progress, Washington D.C., September 13, 2005. Report available online at www.americanprogress.org/autos.

191 two weeks' worth of oil: www.hybridcars.com/blogs/brain/fuel-econ-raised.

192 *So an all-electric:* Joseph Romm and Andrew Frank, "Hybrid Vehicles Gain Traction," *Scientific American* 294, no. 4 (April 2006): 72–79.

192 *8 cents a kilowatt-hour:* Ibid.

193 *The plug-in hybrid will: Reducing California's Petroleum Dependence,* Joint Agency Report, California Energy Commission and California Air Resources Board, Sacramento, August 2003.

194 *overall efficiency:* Romm and Frank, "Hybrid Vehicles"; and Alec Brooks,

"CARB's Fuel Cell Detour on the Road to Zero Emission Vehicles," Evworld
.com, May 2004.

194 *The efficiency of charging:* Ibid.

194 *1,400 gigawatts:* Joseph Romm, "The Car and Fuel of the Future," Report
for the National Commission on Energy Policy, Washington, D.C., June
2005.

194 *less than 400 GW:* Ibid.

195 *overall net emissions:* Lester Lave et al., "The Ethanol Answer to Carbon
Emissions," *Issues in Science and Technology* (Winter 2001). See also Lester
Lave et al., "Life-Cycle Analysis of Alternative Automobile Fuel/Propulsion
Technologies," *Environmental Science and Technology* 34 (2000): 3598–
3605.

195 *Existing oil pipelines:* Michael Bryan, "The Fuels Market—Biofuel Penetration and Barriers to Expansion" (paper presented at the Conference on National Security and Our Dependence on Foreign Oil, CSIS, Washington,
D.C., June 2002), 13–15.

195 *"$2.70 per gallon":* Lave et al., "Ethanol Answer." This calculation includes a
20-cents-a-gallon tax on ethanol. See also Greene and Schafer, *Reducing
Greenhouse Gas Emissions,* 30.

196 *One 2001 analysis:* Lave et al., "Ethanol Answer."

196 *Lee Lynd described:* Personal communications with Lynd.

198 *National Commission:* National Commission on Energy Policy, *Ending the
Energy Stalemate: A Bipartisan Strategy to Meet America's Energy Needs,*
Washington, D.C., 2004.

CHAPTER NINE: THE U.S.-CHINA SUICIDE PACT ON CLIMATE

200 *"The 'international fairness' issue"* and *"We don't need":* Frank Luntz, "Straight
Talk."

200 *2.8 million barrels a day:* Kenneth Lieberthal and Mikkal Herberg, "China's
Search for Energy Security: Implications for U.S. Policy," *NBR Analysis* 17,
no. 1 (April 2006), available online at www.nbr.org/publications/analysis/
pdf/vol17no.1.pdf.

202 *"Perhaps the most":* Richard Benedick, testimony, "The Case of the Montreal Protocol: Science Serving Public Policy," *Hearing on "The Role of
Science in Environmental Policy-Making,* U.S. Senate Committee on Environment and Public Works, 109th Cong., 1st sess., September 28, 2005,
available online at epw.senate.gov/109th/TestimonyBenedick.pdf.

202 *But other uses:* The history in this section is based on Stephen O. Anderson and K. Madhava Sarma, *Protecting the Ozone Layer* (London: Earthscan, 2002); and Benedick, "Case of the Montreal Protocol."

202 *"first unmistakable sign":* Cheryl Silver with Ruth DeFries (for the National Academy of Sciences), *One Earth, One Future: Our Changing the Global Environment* (Washington, D.C.: National Academies, Press, 1990).

202 *"no effect":* Benedick, "Case of the Montreal Protocol."

204 *National Academy of Sciences:* Walter Sullivan, "Scientists Fear Heavy Use of Coal May Bring Adverse Shift in Climate," *New York Times,* July 25, 1977.

204 *"greenhouse gas concentrations":* U.N. Framework Convention on Climate Change, May 1992, full text available online at http://unfccc.int/essential_background/convention/background/items/1349.php.

204 *"Accordingly":* Ibid.

205 *"mandate new commitments":* Byrd-Hagel Resolution, Sen. Res. 98, 105th Cong., 1st sess., *Congressional Record* 143, no. 107 (July 25, 1997): S8113–S8139. Resolution text available online at www.nationalcenter.org/Kyoto Senate.html.

205 *"whereas greenhouse gas":* Ibid.

206 *especially on a per capita basis:* Duncan Austin, José Goldemberg, and Gwen Parker, *Contributions to Climate Change: Are Conventional Metrics Misleading the Debate?* (Washington, D.C.: World Resources Institute, October 1998).

206 *Beijing Energy Efficiency Center:* The discussion of China's energy history is based on Mark D. Levine, "Energy Efficiency in China: Glorious History, Uncertain Future" (remarks at University of California at Berkeley, April 28, 2006); and personal communications with Mark D. Levine.

208 *A 2005 study:* Bin Shui and Robert C. Harriss, "The role of CO_2 Embodiment in U.S.-China Trade," *Energy Policy* (in press).

209 *down to 17 billion tons:* Fiona Harvey and Leora Moldofsky, "U.S. and Australia pledge $128m for climate accord," *Financial Times,* January 12, 2006; and Brian Fisher et al., "Technological Development and Economic Growth, ABARE Research Report" (prepared for the Inaugural Ministerial Meeting of the Asia-Pacific Partnership on Clean Development and Climate, Sydney, January 2006), ABARE, Canberra, available online at www.abare.gov.au/publications_html/climate/climate_06/06_climate.pdf.

CHAPTER TEN: MISSING THE STORY OF THE CENTURY

212 *"In the end":* Maxwell T. Boykoff and Jules M. Boykoff, "Balance as Bias: Global Warming and the U.S. Prestige Press," *Global Environmental Change* 14 (2004): 125–36.

212 *"This is no time":* Edward R. Murrow, *See It Now,* March 9, 1954, available online at http://www.spartacus.schoolnet.co.uk/USAmccarthy.htm.

213 *In November 2005: Meet the Press,* MSNBC, November 20, 2005, transcript available online at www.msnbc.msn.com/id/10042399/.

214 *"If we're unlucky":* Olivia Judson, "Evolution Is in the Air," *New York Times,* November 6, 2005.

215 *"Balance as Bias":* Boykoff and Boykoff, "Balance as Bias." See also www.fair.org/extra/0411/global-warming.html.

216 *To create doubt:* David Michaels, "Doubt Is Their Product," *Scientific American* 292, no. 6 (June 2005): 96–101.

216 *"to spend millions":* John Cushman, "Industrial Group Plans to Battle Climate Treaty," *New York Times,* April 26, 1998.

217 *"suggests that melting":* Juliet Eilperin, "Another Look at Sea Level Rise," *Washington Post,* January 23, 2006.

217 *"is one of many":* Ibid.

217 *"Most scientists agree":* Juliet Eilperin, "Debate on Climate Shifts to Issue of Irreparable Change," *Washington Post,* January 29, 2006.

218 *action on global warming:* The memo is available online at http://desmogblog.com/vampire-memo-reveals-coal-industry-plan-for-massive-propaganda-blitz.

218 *"Coal-burning utilities":* Seth Borenstein, "Utilities Give Warming Skeptic Big Bucks," Associated Press, July 27, 2006. Available online at www.forbes.com/business/feeds/ap/2006/07/27/ap2910768.html.

218 *May 3, 2006,* Washington Post: Doug Struck, "Canada Alters Course on Kyoto," *Washington Post,* May 3, 2006.

219 *"undermines one of":* Juliet Eilperin, "Study Reconciles Data in Measuring Climate Change," *Washington Post,* May 3, 2006.

219 *"over the years":* Wikipedia, s.v. "John Christy," http://en.wikipedia.org/wiki/John_Christy.

219 *"Global warming contrarians":* Richard Kerr, "No Doubt About It, the World Is Warming," *Science* 312, no. 5775 (May 12, 2006): 825.

221 Washington Post Magazine: Joel Achenbach, "What Global Warming?" *Washington Post Magazine,* May 30, 2006.

221 *"We probably won't"*: "2004 U.S. Hurricane Season Among Worst on Record," *National Geographic News,* November 30, 2004, available online at http://news.nationalgeographic.com/news/2004/11/1130_041130_florida_hurricanes_2004_2.html.

221 *Consider a* Washington Post *article:* Lee Hockstader, "Coastal Louisiana Drowning in Gulf," *Washington Post,* July 13, 2003.

222 *Mike Taibbi:* NBC News, "Meltdown," available online at www.dailykos.com/storyonly/2006/1/13/03957/2447; and "January Could Be Warmest on Record in U.S.," Reuters, January 31, 2006, available online at www.msnbc.msn.com/id/11112822/from/RSS/.

223 *"continuous erosion":* Senator Lisa Murkowski, statement, *The Role of Science in Environmental Policy Making,"* Hearing, U.S. Senate Committee on Environment and Public Works, 109th Cong., 1st sess., September 28, 2005, available online at http://epw.senate.gov/hearing_statements.cfm?id=246814.

224 *Climate Extremes Index:* U.S. Climate Extremes Index available online at www.ncdc.noaa.gov/oa/climate/research/cei/cei.html.

224 *They wanted only:* See, for instance, Don Babwin, "Heat Taxes Utilities, Human Endurance," ABC News, August 1, 2006, available online at http://abcnews.go.com/US/Weather/wireStory?id=2261076; and Jennifer Steinhauer, "In California, Heat Is Blamed for 100 Deaths," *New York Times,* July 28, 2006.

225 *"Scientists associate":* Ross Gelbspan, *Boiling Point: How Politicians, Big Oil and Coal, Journalists and Activists, Are Fueling the Climate Crisis—and What We Can Do to Avoid Disaster* (New York: Basic Books, 2004), 79–80.

225 *"Meteorologists are not":* Judith Curry, person communications.

226 *"no prominent national":* Peter Teague, quoted in Michael Shellenberger and Ted Nordhaus, "The Death of Environmentalism."

226 *"Few scientists agree":* Andrew C. Revkin, "Yelling 'Fire' on a Hot Planet," *New York Times,* April 23, 2006.

227 *"The hurricanes we":* Thom Akeman, "Global Warming Behind Record 2005 Storms—U.S. expert," Reuters, April 25, 2006, available online at www.climateark.org/articles/reader.asp?linkid=55586.

227 *"BE WORRIED": Time* 167, no. 14 (April 3, 2006): cover.

228 *"To be fair":* Jerry Adler, "The New Hot Zones," *Newsweek,* April 3, 2006.

CONCLUSION: THE END OF POLITICS

230 *"America is great":* www.bartleby.com/73/829.html.

231 *"In my judgment":* President Bush (remarks at McCormick Place, Chicago, May 2006), available online at www.whitehouse.gov/news/releases/2006/05/20060522-1.html.

232 *"Scientists present":* James Hansen, "The Threat to the Planet," *New York Review of Books*, 53, no. 12 (July 13, 2006): 12–16.

233 *Evangelical Climate Initiative:* Christians and Climate website, www.christiansandclimate.org/statement.

233 *April 2006 news story:* Lauren Morello, "Evangelical Leaders Take Debate to Capitol Hill," *Greenwire*, April 25, 2006.

233 *"government-mandated":* Interfaith Stewardship Alliance website, www.interfaithstewardship.org/pdf/OpenLetter.pdf.

235 *"In nine months":* Doris Kearns Goodwin, *No Ordinary Time: Franklin and Eleanor Roosevelt: The Home Front in World War II* (New York: Simon & Schuster, 1994), 362.

235 *"You see, I told you":* Edward Teller story told in Richard Rhodes, *The Making of the Atomic Bomb* (New York: Simon & Schuster, 1986), 500.

INDEX

Atlantic Ocean (*continued*)
  sea surface temperature of, 34, 39, 43, 50, 51
  see also North Atlantic Ocean
atmosphere:
  aerosol emissions into, 43, 45, 46, 47, 123–24
  circulation patterns in, 37
  greenhouse effect in, 14–15
  greenhouse gas concentrations in, 12, 18–20, *19*, 21, 24, 26, 29, 43, 46, 60–64, 65–66, 73, 80, 87, 92, 94, 104, 153, 155, 201, 204, 206, 211, 231, 234, 236
  in Middle Ages, 65
  ozone layer of, 7, 201, 202–4, 205
  water vapor in, 15, 18, 30, 37, 47, 76
atomic bomb, 235
Austin Energy, 195
Australia, 100, 209
automobiles:
  alternative fuels for, 192, 195–99
  CAFE standards for, 189, 190, 191
  electric drives in, 144, 186
  flexible-fuel, 197–98
  fuel efficiency of, 23, 63, 143–44, 145, 148–49, 188–92, 196, 197, 198, 208, 230–31, 234
  greenhouse gas emissions by, 7–8, 22, 23, 24, 63, 130, 131, 179–81
  hybrid, 23, 143–44, 147–49, 189, 192–95, 196, 197–99, 231, 237
  hydrogen-fuel-cell, 8, 130, 131, 140–41, 142, 143–44, 146, 165, 178, 185–88, 193, 194, 198
  SUVs, 179, 189, 190, 191, 197
avian (bird) flu, 1, 92, 107, 117, 213–14, 223
Avoiding Dangerous Climate Change, 102

"Balance as Bias," 215–16
bark beetles, 56–58
Barnes, Fred, 102
Barrett, Peter, 87
batteries, 147, 192, 194, 199
Beijing Energy Efficiency Center, 206
Benedick, Richard, 202–3
Best Practices Steam Program, 166
biofuels, 23, 55, 63, 170, 192, 195–97, 199
bird (avian) flu, 1, 92, 107, 117, 213–14, 223
bitumen, 181
Bjerknes Centre for Climate Research, 78
Blair, Tony, 100–101, 102, 103, 112, 133–34
Blake, Eric, 47
Bodman, Samuel, 188
Bohr, Niels, 235
Boiling Point (Gelbspan), 225
Bowen, Mark, 110
Boykoff, Jules, 212
Boykoff, Maxwell, 212
Bradford, Peter, 175
Brazil, 13, 72, 198
British Antarctic Survey, 84
British Columbia, 53, 57
Broecker, Wallace, 11, 15, 20, 64
Brooks, David, 99, 113, 116
Bulletin of the American Meteorological Society, 41, 43
Bush, George H. W., 7, 138, 204
Bush, George W., 4, 6, 7, 22, 62, 110, 111, 113, 117–19, 130, 214
  "Advanced Energy Initiative" of, 141, 170
  carbon sequestration issue and, 156–57

sea surface temperatures (SSTs), 34, 35
 hurricane intensity and increase of,
   36–37, 38, 39, 42, 43, 44–46, *45*, 47,
   48, 49, 50, 51, 120, 121, 226–27
 power dissipation index and, 38
 threshold value of, 49
Senate, U.S.:
 environmental policy hearing in, 103,
   122, 124
 Environment and Public Works
   Committee of, 4, 102
 greenhouse gas emissions cap and,
   101–2, 233, 234
 Montreal Protocol and, 202, 203, 205
 UNFCCC (Rio) treaty and, 204, 205
 see also Congress, U.S.; House of
   Representatives, U.S.
sequestration, *see* carbon capture and
   storage
shale oil, 182
Shellenberger, Michael, 129
Siberia, 68–70
Sierra Nevada, 58
Smith, Laurence, 68
smog, 131, 179
snow, 58–59, 76, 77, 78, 80, 222
Socolow, Robert, 22, 24, 63
soils, as carbon sinks, 61, 66, 67–68, 72,
   234
solar power, 23, 63, 155, 167, 169
Southampton, University of, 90
South Atlantic Ocean, 50
space program, 236
Spain, 40, 50, 170
spruce bark beetle, 57
Stanford University, 12, 142
State of Fear (Crichton), 102–3, 107,
   123, 125–28
steam generation, 166, 167
Stephanopoulos, George, 106, 123

Sterman, John, 60
storm surges, 2, 4, 28, 49, 51
stratosphere, 43, 46
sulfate aerosols, 43, 46, 47
sulfur dioxide, 134, 172
sun, 14, 43, 45–46, 47
superconductors, 146
super-hurricanes, 2, 4, 7, 28, 30, 38–39,
   50, 51–52, 64, 91, 151, 236
Sweden, 69, 77
Swetnam, Thomas, 53
switchgrass, 195
Switzerland, 31

Taibbi, Mike, 222
tar sands (oil sands), 181–82
taxes, 134, 138, 189, 190
Teague, Peter, 226
Tech Central Science Foundation,
   142–43
Teller, Edward, 235
temperature, 7, 12–20
 historical change in, 15–17, *16, 19,*
   65–66
 hurricane formation and, 29–30
 measurement of, 25–26
 in Middle Ages, 65
 of oceans, 29, 34–37, 39, 43, 44–46,
   45, 47, 50, 51
 projected rise of, 20, 21, 64, 87, 92, 94,
   204
 records for, 31, 32–33, 41, 52
 in U.S., 2, 51, 54, 56, 57, 58, 92, 94,
   115, 236
 U.S. versus worldwide changes in,
   25–26
 see also sea surface temperatures
Tenet, George, 6
terrorism, 172, 175, 177, 189
Tertzakian, Peter, 180